Anatomy of a Merger

Also by John Stanley Baumgartner
PROJECT MANAGEMENT
THE SILENT WARRIORS

ANATOMY OF A MERGER

—How To Sell Your Company—

by
Robert Q. Parsons
and
John Stanley Baumgartner

PRENTICE-HALL, INC.
Englewood Cliffs, N. J.

Anatomy of a Merger
by Robert Q. Parsons and John Stanley Baumgartner
© 1970 by Robert Q. Parsons and John Stanley Baumgartner

ISBN O-13-035469-1
Library of Congress Catalog Card Number: 70-102102
Printed in the United States of America .   T

Prentice-Hall International, Inc., London
Prentice-Hall of Australia, Pty. Ltd., Sydney
Prentice-Hall of Canada, Ltd., Toronto
Prentice-Hall of India Private Ltd., New Delhi
Prentice-Hall of Japan, Inc., Tokyo

*Dedicated to the professionals who
devote their time and effort to
successful and lasting corporate mergers.*

*And, on a personal note,
to Mary Baumgartner and Nancy Parsons
for their never-ending support.*

CONTENTS

Introduction   9

1. **The Anatomy: An Overview   11**

    The Merger Process   /   The Merger of Electric Products, Inc.   /   Conclusion

2. **Why Sell?   19**

    Reasons for Selling   /   Reasons for Not Selling   /   The Toughest Decision   /   Conclusion

3. **Valuation: "What Price Can I Get?"   27**

    Valuation Yardsticks   /   Effect of Method of Payment   /   Effect of Accounting Aspects   /   Limits on the Buyer's Price   /   Summary

4. **A Peek Inside the Buyer's Tent   39**

    Why Companies Buy   /   The Acquisition Process   /   Pitfalls   /   Financing the Acquisition   /   The Merger-Minded Companies   /   Merger Trends

5. **Earn-Out   55**

    How the Earn-Out Works   /   Effect of Salary on Earn-Out   /   Discounted Cash Flow   /   Definition of Earnings   /   Certificates of Contingent Interest   /   Conclusion

6. **Selecting the Right Buyer   65**

    Selecting a Buyer: The Approach   /   Know Your Buyer and His Proposal   /   A Note on Stock Exchanges   /   Organizational Fit   /   Goals and Planning in Selecting a Buyer   /   Conclusion

7. **Increasing Your Company's Value   79**

    Appearance Value: Merchandising   /   Increasing the Company's Basic Value   /   Magic With Numbers   /   Knowledge of a Company's Value   /   Conclusion

8. **The Merger Broker   97**

    What a Broker Is—and Is Not   /   The Merger Broker: What He Does   /   Fee   /   The Brokerage Contract   /   Summary

9. Legal, Accounting and Other Specialists   115

Selecting the Specialist / Legal Counsel / The CPA / Investment Advisor / Use of Other Specialists / Bringing in the Specialists / Conclusion

10. Merging: Form of the Transaction   123

"Tax-Free" Transactions   /   Taxable Transactions   /   Conclusion

11. Merging: Negotiations   129

Sequence of Negotiations / Preparing for Negotiations / Suggestions on Negotiations   /   Negotiations in Larger Companies   /   Conclusion

12. Merged! The Contract and Closing   139

The Contract   /   The Closing

13. After the Merger   147

Operations   /   Planning and Controls   /   People   /   The Owner Retained   / Conclusion

Appendices

A. Case Histories   153

Leonard McBirnie   /   Arturas Plastics Company—A Divesture   /   Gaines & Berger, Inc.   /   Madlock, Inc.

B. Regulatory Aspects   171

C. Bibliography   175

D. Glossary   183

Index   187

INTRODUCTION

This book is for the owner of a closely-held company that, in a merger, would sell for $250,000 to $10 million. Its purpose is to enable the owner to get full value in the merger transaction, and to provide practical guidance in each step, from the decision to sell to successful integration with an acquiring company.

The owner of a merging company usually faces a well-prepared counterpart in the merger process:

• The potential buyer has gained invaluable experience through previous acquisitions; the seller, on the other hand, usually sells a company only once in his lifetime.

• The buyer has probably made at least a preliminary investigation of the proposed acquisition, and knows the company and the owners much better than they know him.

• The acquirer may have staff specialists whose main function is acquisitions; and in many companies this is a major responsibility of the chief executive as well. The chief executive of a closely-held company, on the other hand, wears a number of hats—but none are labeled Executive in Charge of Selling the Business.

• A merger takes a great amount of executive time; but the time available for making the best sale of his company is very small compared to a buyer's.

• The vast majority of books, articles and case histories on mergers view the transaction from the standpoint of the acquiring company. The buyer therefore has a ready source of knowledge, and can supplement this by seminars designed primarily to assist the acquirer.

• The acquiring company usually has well-defined goals and a plan for accomplishing them, including the place of specific mergers in the plan. By contrast, the seller's goals—personal, financial and business—often are undefined.

In spite of the apparent advantages of the acquirer, however, the chief executive of the selling company has the responsibility for carrying out this one-time activity in a professional way, and for making correct analyses and decisions in this ultimate transaction. As the table of contents indicates, this book will be useful in carrying out this major responsibility.

"What can I get for my company?" This is an owner's foremost question. Some owners sell too low; others have unrealistically high ideas of expected price/earnings multiple. Because of the owner's emotional ties to his business, it is difficult for him to be objective about its worth. And even if he could be objective, establishing a price is subject to negotiation and to various measures of value: potential earnings, book value, market value, earnings history, proprietary features. "What can I get" also depends on the terms proposed and the form of the merger; a seller can, and should, get more in a sale for securities or for cash received in installments than he would in a lump-sum cash deal. Guidelines for getting the full value are outlined in Chapters 3, 5, 6 and 7.

"A good, professional appraisal is good insurance against expensive surprises later on," notes one authority, warning buyers *caveat emptor*.[1] A realistic appraisal is good advice for the seller also, to protect himself against asking too high a price or accepting one that is too low.

Since the book is directed toward the businessman who has knowledge and experience in accounting, legal and investment matters but who is probably not a specialist in any of these, it does not go into detailed matters requiring specialists' expertise. It does, however, provide insight into how each of these functions bears on the merger, and shows how the seller can get the assistance he needs. A particularly valuable specialist, trained as a businessman and skilled in handling the complexities of mergers, is the merger packager. His role and that of other specialists are described in Chapters 8 and 9. These individuals are particularly valuable in view of constantly changing ground rules set forth by government agencies.

In the rising tide of mergers, few companies with growth potential will escape the searching eye of acquisition-minded counterparts.[2] The purpose of this book is to help prepare the owner for a once-in-a-lifetime decision that starts when he receives a letter or a phone call saying, "We would like to see you about a matter of mutual interest. Can we have lunch together one day next week?"

---

[1] Arthur D. Little, Inc., *Mergers and Acquisitions: Planning and Action.* (New York: Financial Executives Research Foundation, 1963), p. 81.

[2] "Sooner or later most corporate managements will have to deal with the merger/acquisition question, either as a buyer or as a seller." *Ibid.*, p. 3.

CHAPTER 1

THE ANATOMY: AN OVERVIEW

A merger is a once-in-a-lifetime wringer through which thousands of closely-held companies pass each year, and out of which flows an increasing number of millionaires. It is a complex jumble of legal and accounting mystique, emotional pangs, government regulations, hard bargaining, qualms, aspirations and discretion. It is a many-sided creature, no two of whom are alike.

When viewed for the first time and as a whole, a merger looks like a confusion of interdependent legal, business, accounting, tax and regulatory aspects. But when dissected, the transaction becomes clear and understandable. It even becomes *logical,* at least as to purpose:

- A producer of aircraft castings wanted to retire. At the same time he was forced to give greater attention to personal matters and, in addition, his industry was undergoing wide-ranging changes. He merged with a company that produced a complementary line of products.

- A family-controlled service company faced increasing labor problems and a declining margin. No capable successors were in sight. Consequently the owners sold their company to a growing acquisition-minded corporation. But they kept the real estate and leased it to the buyer, and as a result made more than they had on company operations.

- Three young executives founded an electronics company. Ten years later, in their late thirties, they found they were bumping their heads against a financial ceiling that stifled their ambitions and their ability to bring out new products. Merging with a diversified company enabled them to give full rein to their capabilities, and gave them access to the financial resources required to fulfill their growth plans.

THE MERGER PROCESS

The merger of a closely-held company usually has commonplace beginnings: growth, which makes it attractive to an acquirer, and the problems that every growing business has. One day the owner comes face to face with the decision of whether to continue to grow or to stand still and eventually be overtaken by competition. He usually faces personal decisions too, such as retirement or how to exercise his full capabilities.

Having fought hard for success, most owners are not about to stand still. But some of the problems will not go away: for instance, how to meet persistent demands for financing, how to diversify the owner's investment, how to get capable people, how to counter increasing pressure from large competitors. Eventually there comes a time when the owner wants to retire and enjoy the fruits of his labor. These factors, a combination of business and personal reasons, are generally involved in the decision to sell. But whatever his reasons, it is a difficult decision because the owner usually feels an emotional attachment to his company.

Price Factors

The main question is, "What can I get for my company?" There is no single yardstick that will give the answer. The price a seller can get is a function of:

- Profit potential. This is the main consideration and is based on the earnings record and the outlook for his industry and his company.

- Price/earnings multiple obtained in his industry.

- Method and terms of payment—whether cash, securities or stock, and whether payable at closing or over a period of time.

- The savings the purchaser will realize from the combined operations.

- The purchaser's limitations: the effect on his earnings per share, the value of the selling company to him, the means of financing available to him, the method of accounting for the merger, and the cost of pumping money, management, time, and so forth into the new acquisition.

Book value may occasionally determine price, and appraisal formulas, market price of a traded stock and other measures of value must be considered. But in the end the price obtainable is a matter of how badly the acquirer wants the company, and how interested the seller is in selling.

The seller will normally fare better as to price and terms, and will better understand the ramifications of the transaction, if he has some knowledge of the acquirer's position: his reasons for buying, how he goes about making an acquisition, what he is looking for, some of the mistakes he has made—and why he is likely to be a very wary cat. The seller should also know something about how a purchaser finances the acquisition, who the merger-minded companies are and trends in the merger surge.

Naturally enough, the seller and the buyer may not agree on the company's potential or on the selling price. To resolve this problem the buyer may suggest an *earn-out*. This is an arrangement whereby if the company does as well in the next several years as the seller claims, he will be rewarded accordingly; but if it does not do as well, his total consideration will be affected. The earn-out is a moment of truth for the seller, but there are a number of pitfalls to be aware of even if his company measures up to his expectations.

There are other ways in which the seller can enhance his company's worth. Presenting a good appearance—merchandising the product—is always worthwhile. And the seller can improve the company's basic operational effectiveness by assessing its strengths and weaknesses and then improving the weak spots. One technique sometimes practiced (but not necessarily recommended) is changing budgets and accounting practices to make profits greater; for instance, cutting back on R&D (research and development) expenses, and changing from heavy expensing to capitalizing equipment over a three-year period. Experienced acquirers are likely to detect superficial improvement, however, through the use of ratios applied to several years' results.

Choosing a Buyer

Usually the owner whose company is sound and has a good growth record will have a choice of buyers to whom he can sell. But how and when to select one is a problem, particularly with regard to what the seller will get in return. If the seller takes cash he runs no particular risk, and can be paid over several years to ease the impact of taxes. On the other hand, equity might prove to be a far better investment, not only from a growth standpoint but also from the standpoint of taxes and avoiding the fact that cash, paid later, is worth less than cash paid "now." But how does the seller know whether he might be getting "Chinese money" in return? How will the acquisition-minded buyer fare in the long run? Different acquirers represent different value, and the seller needs to evaluate the would-be purchaser as well as his offer.

Several very practical problems face the seller in lining up a buyer:

- If the seller himself looks for a buyer, he may give the appearance of "shopping around," with a consequent adverse effect on the desirability of his company as a merger prospect.

- He probably has little or no experience in negotiations and other aspects of the merger process.

- Regardless of how discrete he is, word of a merger may reach the ears of his employees, competitors, customers and suppliers.

- Much time is required, distracting the owner from giving attention to business operations.

- The owner may not know which companies are looking for firms like his, or what they are willing to pay.

For these reasons the seller may ask a professional merger broker (merger packager, merger specialist) to handle the transaction for him.

The seller will also need experienced legal and accounting counsel. "Experienced" means *experienced in mergers*. The company's tax and legal advisors who handle the day-to-day company business ordinarily do not have experience in this increasingly complex field. Thus they are not as well-qualified to handle the merger, although they may have more detailed knowledge of the company. Another advisor whom the seller should consider as part of his team is the investment counselor, who will provide advice on the merits of stock or securities that the purchaser may offer.

The Transaction

The value of these merger professionals becomes apparent during the technical aspects of a merger. For example, a merger may be either "taxable" or "tax-free" with obvious effects on the seller. Where the transaction is tax-free under the Internal Revenue Code, there are three possible forms of reorganization in the sale of one company to another. These are (1) the statutory merger, in which the buyer and seller comply with state statutes regarding the transaction; (2) the stock-for-stock exchange; and (3) the sale of assets. Each of these has advantages and disadvantages to both parties. Whichever form is selected, strict compliance with regulatory requirements is necessary in order to maintain the tax-free nature of the transaction. From an accounting point of view, it becomes significant whether the transaction is a "pooling of interests" or a "purchase," because in the former case the purchaser may be able to offer a higher price than if the deal is a purchase.

Experience pays off also in negotiations. Because of the various trade-off's available to the buyer—price, form of payment, form of the transaction—and the warranties he may ask for, the seller should normally conduct negotiations through a broker or through merger-experienced legal counsel. The importance of experience in dealing with experienced purchasers is indicated by one major acquirer's policy: "You set the price, seller, and we'll set the terms."

The agreement reached by the two parties is reflected in a *Purchase and Sell agreement*. This contract includes the terms and conditions of the agreement and warranties on all representations, such as the seller's contractual obligations and his financial condition. The buyer also makes warranties, such as terms of payment and his status as a corporation.

The last event in the merger transaction is the closing, in which the Purchase and Sell agreement and numerous other documents are exchanged. This usually takes place thirty to sixty days after the Purchase and Sell agreement is approved by the two parties. Finally, at closure the seller gets his certified check or other consideration for selling his company.

A brief summary of an actual merger will serve to bring these aspects of the transaction into focus.

THE MERGER OF ELECTRIC PRODUCTS, INC.

Clint Jordan's problem was success, and taxes. Jordan really wanted to run a small business and keep a personal hand in product development. But he woke up one day to find that he was operating a 150-man organization and that he was spending most of his time on Electric Products' administrative problems and marketing. About the same time, both his accountant and his attorney informed him that he faced a heavy tax penalty on excess accumulated dividends. The question facing Electric Products' owner-president was what to do about it.

Company Background

Electric Products had done very well after a rocky beginning the first few years. In fact, the company had accumulated more than $250,000 in cash in recent years, and Jordan's attorney and CPA advised him that he could be socked for almost $50,000 in taxes (27 1/2 per cent on the second $100,000, and 38 1/2 per cent on the balance). This was one of Jordan's problems.

But success also raised some fundamental personal problems. Jordan was basically a mechanical engineer who had made money after World War II in construction and real estate. But a desire to tinker and develop better electrical products used in construction led him to spend more and more time at this avocation. Finally he plunked down $10,000 of his savings and started Electric Products. For tax purposes he set up a production company and a marketing firm that bore the Electric Products name.

Ironing out bugs in the basic product and in the molds required to produce them caused a heavy drain on finances in the first two years. But then, over the next several years, the basic merit of his products became so well recognized that his company gained a major share of the market. By the time of his discussions with his CPA and his attorney regarding the excess accumulated dividends, the company had captured 90 per cent of the Western market for these specialized products.

Jordan's main problem—other than taxes—was that he was spending more time on general administration and marketing, which he did not particularly like, and less time on product development, where his own talents and the fundamental strength of the company lay.

Electric Products' president faced these decisions: whether to bite the bullet and pay taxes on the accumulated earnings, or invest the funds in some other way; whether to stagnate at his company's present level or continue to grow; whether to devote less time to product development in order to give more time to managing the business. None of these choices appealed to him.

Another problem began to creep up also. Jordan drew $45,000 in annual income; charged to the business a company car, a plane he piloted himself, and other expenses; and he was living comfortably. But he began to give increasing thought to how he could someday get his money out of the company.

A Decision to Buy

Jordan made a decision: He would buy a company to utilize the excess cash. It was a tough decision because he realized he would necessarily be spending more time on management problems and less on product development.

Jordan contacted a merger specialist, Roy Peterson, to discuss buying a firm that would complement his own and not require too much management attention.

The specialist's first question was, "What do you really want to do? What are your objectives?"

"To buy a small company that complements Electric Products."

"But that'll take time and attention away from your own company. With several new products already in your patent attorney's hands, it looks to me as though you'll have your hands full just keeping up with Electric Products."

"Yes, but I've got to do something with this excess cash, and soon." Later in the discussion Jordan raised another question. "Say, how does a guy ever get his money out of his company?"

The question remained unanswered while the merger specialist searched out several possible acquisitions. None, however, was what Jordan was looking for; in fact, Jordan began to recognize that he did not really want to buy another company.

The Decision to Sell

One day some months later the specialist said, "Clint, it's obvious to me that you'll be much happier if you *sold* your company, rather than buy another one. You have far more interest in developing products and in dabbling in real estate than in staffing and running an industrial organization. You like a small operation where you can make the decisions and do practically everything yourself, and Electric Products isn't ever going to be the way it used to be. Have you ever thought of selling the company?"

Actually Jordan had thought of this possibility before; he had turned away a number of would-be acquirers. He thought about the question a few moments. "All right, find a buyer for me. How much do you think I can get for it?"

"I don't know. But after I make an analysis I can give you a pretty good idea."

Peterson made an analysis of the company's operations, profit and assets, and worked with the company's CPA to make up a pro forma balance sheet and operating statement. They recast these statements in the form normally used by major corporations; for instance, they eliminated the company airplane and one of Jordan's salaries (he drew a salary from both the sales company and the manufacturing company). The resulting profit picture, based on a projection from the previous five years' operations, showed a very attractive profit growth. The potential extrapolated from this record and the corresponding price/earnings multiple indicated a selling price of about $2 million. Jordan and the specialist set this as the asking price, in cash or negotiable securities.

The Merger

As a result of contacts made by the specialist, without revealing the name of Electric Products, Inc., several companies were interested in discussing merger prospects. The most interested proved to be a New Jersey-based producer of electrical equipment for consumer and industrial uses, Brighton Industries. Brighton's management looked at the profile prepared by Peterson and analyzed how the company would fit in with their own goals and operations. It looked like a natural fit. As a result, Brighton's president, Fred Kohler, and his vice-president for manufacturing came to discuss the transaction and negotiate a merger. They were well aware of the $2 million asking price, and had instructed their lawyer to draw up a Purchase and Sell agreement in anticipation of finalizing negotiations.

The shocker came when Kohler said, "We think Electric Products will fit in very well with our company. We're prepared to offer you $1.7 million for it." This almost killed the deal on the spot. The merger specialist and Jordan were convinced the company was worth the asking price, particularly since it fitted in so well with Brighton's other lines. After considerable discussion, Peterson made a suggestion: $1,850,000 in cash, with a five-year consulting agreement with Jordan at $30,000 per year. This was finally accepted by both sides.

What should then have been a relatively smooth process of drawing up the legal documents was another matter that almost killed the deal. The buyer's attorney was a junior man, working with minimum guidance under a senior one. Being overly cautious, he inserted almost *twenty pages* of warranties binding the seller. When Jordan read the warranties he was being asked to sign, he exploded. "To hell with it! What are *they* warranting?" At this point, with the ball in his court, Peterson had the dual role of negotiator and educator. He explained to Jordan that their certified check was primarily Brighton's warranty. And he explained that because purchasers have so often in the past been "burned" from not having investigated thoroughly (and in many cases they *cannot* investigate thoroughly, without disclosing to employees and others why they are in the seller's facilities), it is customary for the purchaser to require warranties as to the selling company's status. At the same time, however, Peterson pointed out to the purchaser's lawyer that some of the warranties asked for were unreasonable. These were then deleted from the draft agreement.

Another crisis occurred when the buyer was misinformed regarding reported changes that he attributed to the seller's lawyer, but which were in fact traceable to a misunderstanding on the part of his own attorney. A more serious problem arose when the buyer's attorneys insisted that $500,000 of the $1,850,000 cash price be set aside in escrow for three years to protect the buyer against a failure on the warranties. This was reduced by agreement to $300,000 in escrow to be released within a year of closing. This was acceptable to Jordan because it would give him most of the cash required to pay his capital gains income tax. Another problem arose when it was discovered that the inventory included some defective parts. This was overcome by agreeing that any deficiencies would be charged against the warranty. (This eventually resulted in a charge-back of $12,000 against the amount in escrow.)

After all these hurdles, the day came for closing. Because of the time and attention given to details prior to closing, the closing itself was a very smooth transaction. Jordan received a certified check for $1,550,000, a consulting contract at $30,000 for five years and $300,000 placed in escrow payable within a year. But he also received something equally important to himself: freedom to exercise his talents in inventing and developing new products.

CONCLUSION

No two mergers are alike, any more than two companies are alike. The anatomy of a merger bears similarities from one transaction to another, however, and in some ways Clint Jordan's merger resembled many other mergers of closely-held companies.

- His reasons for selling were a combination of business and personal factors (tax implications and personal desire to create new products).
- The price he could get was based on a combination of historical record, profit projections, form of payment and value to the acquirer.
- Recognizing his own limitations in effecting a merger, he engaged the help of a merger specialist.
- He was able to choose a purchaser to whom his company was particularly attractive.
- He had excellent legal counsel, experienced in mergers. In fact, his own counsel proved to be more capable than the purchaser's.
- Because of his advisors' experience, Jordan was well represented in negotiations. Thus the closure went very smoothly.

Jordan's merger differed from some in that there was no earn-out feature; his company was already saleable—no "dressing up" was required; in selling his company's stock he elected to receive cash rather than equities or securities of his acquirer (in addition, Brighton was a family-owned company), and because of this he did not need investment counsel.

As in most mergers, Jordan retained an active interest in the company, in this case as a consultant. And, particularly important to him, the transaction left him with the capital and the time for what he particularly wanted to do.

CHAPTER 2

WHY SELL?

Why do the owners of closely-held businesses sell? The reasons run a broad gamut, as these quotes indicate:[1]

"Our family investment in [the] company was too high in relation to income or other holdings."

"Pay the blasted income taxes."

"Lack of competent management."

"Our business is very competitive and while we made profits every year, we could not seem to generate sufficient profits to keep improving plant to compete with large competitors."

"We had an experience in a gift tax valuation of our stock that alarmed us as to coming estate taxes."

"To enjoy my 'estate' while still living."

"Due to lack of working capital to expand to meet competition."

"I did not seek a buyer. . . . Their offer to me was attractive and the responsibility of sole ownership in a rapidly expanding operation caused me some concern, so I decided to sell."

"A chance to sell stock. . . at a much higher price than book value with a result-ing profit to the stockholders."

"After the first Union meeting, I knew that I would never be able to operate a plant with the type of Union management we have in this country."

"Age of management primary factor in sale—no new blood to take over. Present generation unwilling to put in hours to make a success of business."

[1] Chelcie C. Bosland, *Estate Tax Valuation in the Sale or Merger of Small Firms.* (New York: Simmons-Boardman, 1963), pp. 55–65.

"My partner died and it gave me an opportunity to settle with his heirs."

"My father was a very successful manufacturer with a profitable earning record (one of the best). He died and taxes made us lose control of the company and also took all of the money from us that was usually used to expand the business. It 'hit' at the heart of us—we sold out."

Another and different reason, however, is the case where a young, aggressive executive wants to merge in order to have a key position in a larger, aggressive company.

Generally no single reason is in itself enough to induce owners of a closely-held company to sell; usually there are a combination of business and personal reasons. In toto, these amount to either the expectation of gain, or fear of loss in continuing the business.[2]

Not all of the reasons for selling are revealed to buyers (as they have discovered through painful experience), and the seller himself may not fully have probed his own purposes. Sometimes there are good reasons for selling that are not apparent to others, including employees. In one case a seller accepted $3.5 million for a company that had almost $3 million in the bank. Key employees and others were mystified and even bitter about the merger. What these observers did not realize is that by leaving the cash in the bank, the owner could have been taxed as much as 70 per cent on undistributed dividends; if he had distributed these funds, he could have been taxed up to 90 per cent in income taxes. Accordingly, it made good sense to the owner to sell.

The decision to sell or not to sell is often the toughest decision an owner makes. He has strong emotional ties to his "baby," particularly if he has raised it from a gleam in the eye to a solid business with a record of good results.

This chapter highlights the major points to consider in the sell/no sell decision.

REASONS FOR SELLING

The overt reasons for selling are usually a combination of business and personal reasons. Business reasons include the need for financing, operating economies, ability to grow faster because of a buyer's strengths (capital, market outlets, plant, R/D capabilities) and deteriorating sales or earnings record. In addition, there may be severe competition, lack of management continuity, labor and union problems, narrow product line or narrow market base, high price of automation, complexities of doing business and, naturally enough, an attractive offer from a buyer. Personal reasons include the valuation factor (uncertainties of the value of closely-held stock), investment diversification and lack of liquid assets, estate tax considerations, desire for capital gains, wish to retire, dissent or lack of interest among the owners, desire to

---

[2] One authority notes that such factors as an outmoded product line, inability to cope with technological advances, fear that key people might leave, fear of losing a lifetime's accumulated capital and fear about the future are present in most decisions to sell, although these reasons are not often expressed in this way. See Myles L. Mace and George C. Montgomery, Jr., *Management Problems of Corporate Acquisitions* (Boston: Graduate School of Business Administration, Harvard University, 1962), p. 33.

become an officer in a larger company through merger and greater challenge for capable management. Some of the more common reasons are discussed in the following sections.

Tax Considerations

Federal taxes take 52 per cent of net corporate income over $25,000 and up to 91 per cent in personal income tax on income above $200,000; estate taxes are as high as 77 per cent on estates over $10 million. State taxes further reduce the amount available to the company and to its owners. The owner-manager of a closely-held company begins to wonder, "Why should I knock myself out when the company has to make four or five bucks for every one I get?"

High income taxes tend to slow the potential growth of smaller corporations (more so than large ones) because retained earnings, the small company's main source of financing, are reduced by half in income taxes, limiting the company's borrowing power and also crimping the motivation to keep on growing as a corporate entity.

Estate taxation is probably a greater factor in selling or merging, however. Consider a company in which three major owners invest $300,000. The firm grows through the years under their guidance and has a value to a potential acquirer of $6.3 million. If sold at this price, each owner would receive one-third of the $6 million capital gain, or $2 million. At the capital gains tax rate of 25 per cent, he would pay $500,000 in taxes, but would still have a $1.5 million estate. If he were to receive stock in the acquiring company of value equal to his $2 million, however, he would pay *no* tax at the time of the transaction, and could later take tax-favorable action with regard to it.

Suppose, however, that one of the three owners in this situation dies. In order to pay estate taxes, his heirs, who do not have substantial assets, try to sell his interest to the other two owners. Will they—or the company treasury—have enough liquidity to effect the purchase? In view of the various ways of placing a value on the company, how much is his share worth? Won't a purchase of his equity reduce the other owners' liquidity, which might have been used for expansion and consequent increase in the value of the company? It is worth noting that tax appraisers may have a higher (sometimes much higher) evaluation of the company than do the owner-survivors. For owners facing this kind of situation, a merger may provide a clear-cut and perhaps profitable solution.[3]

Investment Diversification

Desire to diversify his personal investments is a major consideration for most owners whose assets are tied up in the company. At some point in his career, having plowed

---

[3] There are alternate ways of dealing with this problem: purchase of stock by the business or survivors, personal life insurance, gifts before death, public sale of stock before death, gifts to charitable institutions, maximum use of the marital deduction. But each has attendant problems. See Bosland, *op. cit.*

most of his earnings back into the ever-increasing need of his company for finances, the owner begins to wonder, "How do I get my dough out of the company? And how secure is my capital, in view of the flux going on in my industry—market and product diversification, and the big ones gobbling up the small ones?" He may well conclude that a sale for cash or stock in a growing, diversified company might yield sufficient income and be a sounder investment than keeping his eggs in the company basket.

Need For Capital

Ever-increasing accounts receivable, inventories, payroll and need for more space and equipment combine to make more demands than can be met by the generation of profit in many growing businesses. Restrictions on use of loans, as well as amount, may limit this means of financing.

Accordingly, financing is a prime reason for merging in the case of both the successful and the not-so-successful company. A typical example of the need for financing in a successful company was the manufacturer of lighting fixtures whose business had grown at a rate of 25 per cent for seven successive years. All of the firm's earnings and borrowing power had gone into the expansion. At that point in the company's growth, the owner discovered that forthcoming changes in his industry would make his tooling completely obsolete. Financing the tooling required for him to remain competitive was beyond his strained financial capacity. He decided to merge, and subsequently was acquired by a company listed on the Big Board in return for a management contract, stock in the acquiring company and assurance of working capital necessary for expansion. Two years later, as a result of the parent company financing the necessary tooling, he was able not only to hold his own in the market but to gain appreciably over his competitors.

In the case of a manufacturer of fire extinguishers, the company's position had steadily declined. Although it was developing some excellent product ideas, it was unable to bring these to profitable fruition. The company merged with "Automatic" Sprinkler and received a quarter of a million dollars in financing. Subsequently it brought out seventeen new products and headed back to prominence in its field.

Product Development

As the market life of products gets shorter and shorter, and as technology advances at a faster and faster pace, concentration on new products requires both financing, as in the preceding cases, and know-how. The owner-manager who started his company with abundant know-how may discover that he has become technologically obsolescent. This happens about every five years in high-technology industries, and even the biggest companies bump heads with the obsolescence factor.

A giant computer manufacturer, for instance, decided to go into "platform" development business in aerospace guidance and controls systems. After being in the business for two or three years, the company made the unhappy discovery that it was a year or two behind in technology and losing ground; in addition, the field was over-crowded with hungry, experienced competitors. The program was packed off to the company's skeleton closet. The funds, personnel, equipment and so forth for the program were assimilated into the activities of other departments, the equivalent to the merger of a closely-held company.

Abundance of technological talent in a closely-held company, on the other hand, is a commodity highly sought by acquisition-minded companies.

Retirement and Lack of Succession

Neither chance nor design provides a successor in many closely-held companies; and as affluence and leisure time of others increase, it is natural that a hard-working owner-manager might want to retire sooner than his father and his grandfather did. The axiom that every key individual should have a back-up man rings through loud and clear one day when the owners conclude that they should consider selling because they do not have suitable successors for top management. They may also have to be prepared for a lower selling price because of this thinness in management.

An unusually fortunate company in regard to management succession is a Western producer of sports car accessories and aircraft parts. The company was founded in 1946 by four partners, including a father-son team. The firm weathered various financial crises and grew to a sales volume of about $5 million. At this point the elder three of the four partners wanted to retire from active responsibilities. The fourth partner, a generation younger than the other three, was highly capable, had more than twenty years' experience in the firm, and still had the best years of his career in which to carry the company to a high plateau of earnings and potential. In less fortunate companies, where owners' sons have chosen other careers or are not ready for heavy responsibility, or where management personnel do not have the business background and dedication required, the *position gap* (the gap between what the position requires and the capabilities of an individual to fill it) leads to a decision to sell.

In some cases a family successor has taken the reins but finds that he does not like being the chief executive. This was the case where the third-generation head of a prominent business equipment company merged with a fast-growing diversified company.

Similarly, a second or third generation son sometimes finds he is responsible for running the operation and has perhaps a quarter of the stock, the other shareholders being sisters or widows. In effect he is responsible for their estates. Because of this he may elect to merge with a publicly-held company to relieve himself of this responsibility and to give them a more liquid position.

Competition

Formidable competition frequently leads to the conclusion "maybe we should sell." On the other hand, these same formidable competitiors sometimes see that the route to faster growth is through a selected merger partner who can provide the ready marketing channels, financial resources and management controls required to achieve competitive advantages.

An Attractive Offer

In this era of the seller's market, a decision to sell may be triggered by an unusually attractive offer. In addition to receiving value in return for the sale of their company, the sellers may be able to remain active in their same positions in the new subsidiary operation, with satisfactory remuneration and an active career as well as a voice in operations of the parent company. The owners need to know, however, what constitutes "an attractive offer"; that is, how to value their company and how to evaluate the stock that might be received from an acquirer.

The Silent Partner Situation

Sometimes the reason for selling is the situation where two partners, one of whom initially backed the enterprise, have seen the company grow and prosper. They split the rewards equally, but one works particularly hard. In a typical example, a fifteen-year-old firm was doing $5 million in annual sales, and both the silent partner (who had put up $15,000 to start the company) and the operating head drew $75,000 annually, excluding dividends. The active partner was working seventy to eighty hours a week for the company's continuing success, but was getting only half of the financial rewards. He wanted to sell and, with his financial partner, was able to effect a satisfactory merger in which each received stock in a publicly-traded company, and the working partner got a five-year contract at $100,000 a year to continue to manage the company.

REASONS FOR *NOT* SELLING

Some publications consider the questions "why buy" and "why sell" as if there were only positive reasons for each. In the case of selling, at least, there clearly are factors to consider on the negative side of the question, because of the finality of the decision. Some of the reasons for *not* selling are these:

Loss of control. The acquiring company controls the purse strings and has "command authority" over the previously independent operation. The individual who sells a company has to realize that in the event of a sale or merger he is no longer the ultimate decision maker. In numerous instances the owner stays on under contract and continues to manage the company, but often he is unable to adjust to the fact that he has to go to someone else for the final decision on certain matters. This is one of the prime reasons that an individual may not be happy after the sale of his business.

Loss of identity. The name of the acquired company may be lost in the merger, or at a later time due to subsequent mergers of the acquiring company.

The "unknown." The seller might be joining forces with a largely unknown partner, without knowing what might happen to valued employees and without knowing the "real" value of the buyer's stock received.

Solving its problems. The company may be able to solve its own problems by getting better management; getting more effective marketing personnel, products and outlets; making a thorough search for possible sources of financing; or by acquiring smaller companies who have complementary strengths.

Other alternatives. Rather than merging with another company, the owners might consider going to public to raise money, or selling assets or a product line. Another possibility is indicated by the brief encounter between a Washington-based nuclear consulting firm and a would-be acquirer, who asked, "Are you interested in selling your company?" The reply: "No, but we've been looking you over. How about selling yours to us?"

Loss of small business qualification. A company might grow, by virtue of a merger, to the extent that it loses its favored position as a small business supplier to the government.

Specialized service. Because of its smaller size and flexibility, and the fact that it can provide responsive service and products to a particular market, a company may elect not to sell because it would lose these advantages.

Not a good-enough offer. For various reasons, offers to buy may be unacceptable. Mace and Montgomery cite the case of a West Coast company that received an attractive offer from a company that was well-known and highly respected by the principals of the Western firm.[4] The merger had all the features that make for success, including the right "chemistry" and "fit." But the company's management was competent, and there were successors for all top management positions; its market was varied without undue reliance on any single customer; manufacturing facilities were adequate, sufficient capital was available, there was no threat of product obsolescence; R&D was producing profitable new products, and key management and technical personnel were young and wanted to go it on their own. In other words, *there were no valid business reasons* for merging. The principals decided not to sell.

In deciding whether to sell, an owner should ask himself these key questions:

Are there valid *business* reasons for selling the company? Can problems and the reasons for selling be overcome?

Are there valid *personal* reasons for selling the company? Can personal problems be overcome?

THE TOUGHEST DECISION

The question whether to sell his business is particularly difficult not because of whatever the owner's logic dictates, but because of the years and perhaps generations of

---

[4] Mace and Montgomery, *op. cit.,* p. 28.

sweat and emotion that have gone into the business. This is the toughest decision. It is the kind that an otherwise strong, practical businessman finally faces—and puts out of his mind. Faces again and puts out of his mind again. This can go on for months.

One way for the owner to find out for himself whether this is the step he really wants is to prepare a paper model such as an acquirer would like to see, and include these items of information on his company:

- History of the company

- Information on owners and distribution of ownership

- Information on marketing, production, R&D, general management and finance and control, citing strengths, weaknesses and problem areas.

- Rate of growth

- Reasons for selling

- What the seller expects from a buyer, in terms of what he thinks or expects the buyer to do for his operation

- An idea of what the seller expects to get (cash/stock/other) for his company, and the management position he wants, if any, in the merged organization

- Operating statements and balance sheets for the previous five years, and projections three years into the future

If the owner does, in fact, go through with preparing this information (or has a broker do it), he is probably ready for a merger; and the model of his company will contain the information that a potential buyer will ask for in his preliminary investigation. If he finds the exercise is not worth his time and effort, he is probably not ready to sell his company. In either case, thinking about the business in this way will probably result in his taking action to make it a stronger, more profitable operation.

CONCLUSION

The answer to "why sell" is often not easy, for merging is an unusual event in the life of the company and the lives of its owners. It is particularly difficult because of their emotional attachment to the company. The unknowns in a merger can be stripped away, however, by giving the same professional thought and attention to this aspect of his business that the owner would to any other. He may find that his "why sell" questions are easily answered when he gets a response to the key one: "What price can I get?"

CHAPTER 3

VALUATION: "WHAT PRICE CAN I GET?"

Even before he has determined that he wants to sell, the owner wants to know, "What price can I get?" This is understandable, for if the price is right he might change a "no sell" decision to "sell."

The merger urge affects so many businesses that the owner of a closely-held company probably knows personally of instances where another owner sold for an unexpectedly favorable offer, and of other instances where offers were refused because they were considered too low. What he may only vaguely realize, however, is that sometimes companies are sold at *too low* a price, often without the seller ever finding out.

A banker cites an example where the owner of a company stated in no uncertain terms, "I won't take less than ten to twelve times earnings!" His after-tax earnings were about $150,000; indicated book net worth was $1 million. He was offered $2 million and accepted promptly. What he still has not found out is that the buyer, who had a current appraisal made, learned that *the seller's real assets alone had a $2.5 million current value.* The seller was looking at only one measure of valuation. The buyer got a going business and $2.5 million in real estate for an outlay of $2 million.

Occasionally even the buyer will acknowledge that a seller has received less than his company was worth. In contrast to the seller mentioned in the preceding paragraph, however, many sellers overestimate the price they can get. Two shocks are in store for the seller of a closely-held company. One is the day he realizes the company is no longer his. The other, pertaining to price, is the day he finds out what his company is really worth.

A seller can get unrealistically high visions of sugarplums by comparing his potential price/earnings ratio with what he reads that others have received.[1] He might be overlooking the fact that the average stock in the Dow Jones Industrial Index can be bought for about sixteen times earnings.

Price is determined primarily by what the buyer thinks he can make from the company. He is not willing to pay the seller for what the buyer contributes—for instance, operating capital—to make this expected profit. For the seller, finding out the price can be a trip from the depths to the heights to the subbasement as he in turn forms an estimate of value, thinks his price should be much higher because he has been holding down taxable earnings, and then learns what he can *really* get for his company.

It is understandable that an owner, having emotional attachments, finds it difficult to value his company. After all, experienced acquirers who have no such attachment also have a difficult time in evaluating an acquisition.[2]

Business Week comments: "The issues in a merger are involved and complex. But they all boil down to a question that has bothered economists for centuries: What is value, and how can it be measured?"[3]

In spite of difficulties in measuring value, there are criteria for valuation that provide guidance to the uppermost question in the owner's mind: What price can I get? This chapter presents ways of determining value and price, and how to get the full price due.

VALUATION YARDSTICKS

Most acquisition-minded companies, investment bankers, merger brokers, accountants and other experienced individuals recognize *earnings potential* as the primary criteria for merger valuation. But they are quick to point out that other measures must also be considered, since no two cases are the same and since each type of measurement has some bearing on valuation. *Book value* is an obvious measure, and one to be considered particularly in companies where the price based on a multiple of earnings is below book. But for most growth companies subject to acquisition, the use of book value may cause a valuation below what the acquirer would actually be willing to pay.

[1] Ninety-four times earnings in the 1969 merger of Scientific Data Systems with Xerox; and a fantastic 371 times price/earnings valuation recounted in "The Fastest Richest Texan Ever," Arthur M. Louis, *Fortune,* November 1968, p. 168.

[2] Companies that have made five or more acquisitions have a significantly higher percentage of "successful" acquisitions than those with less experience. To determine what is a successful merger, one must apply the twin yardsticks of time and the question, "Would I do it over again?" Where the acquiring company has the muscle, talent and time, it can turn an unsuccess into a success. Consequently, in many mergers the answer, five years after the merger, is, "Yes, we would have done it again, but we would have done it differently." In connection with this, one analyst notes that "the key to a successful or unsuccessful merger policy is the price paid for the acquisition." Eamon M. Kelly, *The Profitability of Growth Through Mergers* (University Park, Pa.: Pennsylvania State University Press, 1967), pp. 70-71.

[3] *Business Week,* Nov. 23, 1968, p. 142.

The main purpose of book value, in fact, may be to show whether or not relatively conservative accounting practices have been followed. *Dividend-paying capability* is a consideration, because the ability to generate cash throw-off is an indication of the ability to reward owners for their risks. In publicly-traded stocks, *market prices* indicate valuation; in fact, as noted in Chapter 2, this valuation established by the market is one reason for going public. Other measures of valuation are *patent rights* (as in the case of a company with $600,000 in sales and six employees, which sold for $1.5 million because of proprietary product); *gaining management and technical personnel* who have demonstrated measurable profit-making potential; *adding sales outlets or facilities* of value to a potential acquirer.

The owner should ask himself two primary questions: What do I have that is of value? Of value to whom?

Before looking at the various measures of value, it might be well first to consider the five P's that bear on each measure: PERSONNEL, PRODUCT, PLANT, POTENTIAL, PROFIT. Some sophisticated buyers rate each of these factors on a ten-point scale, for a possible price/earnings multiple of fifty. For factors that rate lower on the scale, the price/earnings evaluation is marked lower accordingly. A seller might consider rating his own company on a similar basis.

Personnel

Most experienced buyers, such as the chief executives of conglomerates, say that what they really are acquiring is a going organization. Generally they want a company that is fully staffed, with a general manager and able functional heads; and, since it takes three to five years to develop a good operating team, they want assurance that these key people will stay on the job. The characteristics these buyers are looking for are cited by Willard F. Rockwell, Jr., Chairman of North American Rockwell, who quotes a vice-president of an investment bank: "Three common characteristics of able executives are unusually high degree of motivation, energy, and intelligence. Not all successful executives have all three, but few have less than two."[4] Some high-multiple acquisitions have been made primarily to acquire a strong management team, or even a single outstanding young executive.

The personnel factor is also the basis that some acquiring companies use in setting a minimum size. One company, for example, has set a minimum of $5 million in sales, because an acquisition at this level will be staffed in all major functions: It will have a comptroller, a production manager, a chief engineer, a marketing manager, a quality control head, a personnel manager and a competent general manager. In smaller companies the chief executive wears many of these hats.

The organization factor carries as much weight as the plant or product in determining the overall price/earnings multiplier.

---

[4] Willard F. Rockwell, Jr., "How to Acquire a Company," *Harvard Business Review,* September-October 1968, p. 124, quoting John R. Shad, Vice President of E. F. Hutton & Co., "How Investment Bankers Appraise Corporations," *The Commercial and Financial Chronicle,* Aug. 2, 1962, p. 15.

Product

The company that has a proprietary or patented product rates high in value to a prospective buyer. An extremely high rating is often merited by a patented product that has not been exposed to national sales.

Other product factors relating to company value are products being developed that dovetail into a prospective buyer's requirements; products for which extensive tooling has been accomplished; a recognized brand or trade name that has been established over a long period.

The value of customer acceptance as a multiplier is indicated by the product position of a nationally-known producer of consumer items. A chain of stores that accounted for 5 per cent of the producer's total sales wanted a more favorable discount. The producer refused to bow to the chain's demands for a change in discount policy, and their products were removed from the chain's retail markets. Within thirty days consumer demand was so great that the chain reinstated all of the products, at the producer's regular discounts.

Plant

Since most buyers are looking for growth, it is axiomatic that most sellers should put their growth foot forward in terms of sales and profit forecasts. This raises questions of production. Does the selling company have the production capacity to meet its sales forecasts? What is the condition and age of the equipment—will it have to be replaced with less worn and more efficient equipment? Is the equipment for a special purpose or does it have a more general application? Arrangement of a plant is significant; a Chicago plastics plant in a multistory building was an attractive acquisition partly because gravity flow of raw material made its operations particularly efficient.

A plant with a capacity of $8 to $10 million, but with current production of $3 million, is more attractive than a plant that is operating at its peak and will have to be enlarged or expanded in order to realize the projected sales growth. An ironic situation occurred where a plant was operating at peak efficiency and at the limit of its capacity; a buyer turned it down because he could not see how he could improve earnings or increase production without immediate capital expenditure.

Highly sophisticated equipment, such as that required to produce integrated circuits, naturally rates higher than simple equipment that requires a high labor expenditure or has limited growth potential.

The appearance of the plant is carefully noted by the specialist brought in by the acquiring company. Material flow, cleanliness, condition of the machines and general efficiency of the operation are weighted in rating the plant.

Potential

The company's historical rate of growth as well as the development of its industry have a marked effect on the value of the company. If the company is in a growth industry (growing 15 per cent or more per year) and if the company itself is increasing this fast or faster, it will rate high in attractiveness to a merger partner. Where the rate of growth (sales and profits) is around 10 per cent, or comparable to the growth of the gross national product, the price/earnings multiple will be considerably less. The reason for the growth is significant also; if it is due to aggressive management, as well as happening to be in the right field at the right time (for instance, in a growth field such as leisure products or computers), the long-run potential will be marked higher accordingly. Growth in earnings, rather than in sales, is the key factor in evaluating potential.

Profit

This is the base for determining the price/earnings multiple, to which the other P's are applied. A well-run, privately-held company will use every legal means *not* to show profits (double-declining balance on depreciation, expensing rather than capitalizing, heavy contributions to bonus and profit-sharing plans, and so forth). In preparing a company for sale, however, the profit-and-loss sheet and the balance sheet should be recast in the same form as an acquiring company's statements. In one case a job shop developed a technique for heat-treating disc-type memory units for a computer manufacturer. Its profit was only $9,000. But this profit was recast by an experienced merger-packager to show an equivalent net of $50,000; the buyer paid twenty times earnings based on the latter figure.

A company having a higher *gross* margin (for instance, 40 per cent versus 19.5 per cent) bears a higher price/earnings multiple because, whatever the net, sophisticated buyers know they can bring overhead and general and administrative expenses (G&A) in line with industry averages. Some acquirers whip their acquisitions' overhead costs into shape surprisingly fast.

Because of the importance of profit and potential in determining price, a more detailed discussion on future earnings is in order.

Future Earnings

The seller's estimate of future earnings considers, first, what his future earnings would be as an independent organization; and, second, (when he has established actual or possible acquirers) what his future earnings are likely to be in view of the improved operations to be achieved after the two companies have merged.

1. *Past earnings.* Most projections of earnings start with a review of past earnings, typically for the previous five years. To be realistic, the owner should isolate conditions that may have had an unusual (for better or worse) result on earnings, such as his main competitor being hit with a long, expensive strike, or his company being in a favored supplier category because a large customer happens to be managed by his brother-in-law. To the extent possible, he should adjust his income statements to correspond to those of a publicly-held corporation (that is, adjusting for expenses incurred for his wife's Cadillac; adding to his own salary if he has reduced it for tax reasons). The owner should also analyze the reasons for his earnings (because this is

what a prospective buyer will do). Has his market penetration increased? Why? Is his profit due to abnormally reducing R&D and sales expenses? Which product lines contributed most to his success, and can some be promoted more and others deleted in the future?

2. *Future earnings.* The owner should then develop forecasts of profit and sales, if these are well-enough thought out to be accepted by a buyer's planning personnel, taking into account industry growth and competition, plans for expansion, new products and anticipated problems. An aggressive purchaser will analyze the selling company in six major areas, and will make a thorough analysis in each of them. These are: *competition,* present and projected, in each product line; the seller's *market* in each of his present and future product lines; past and forecast *performance,* in sales volume and share of the market; *future requirements* in terms of finances and personnel; *financial analysis,* with particular attention to historical and projected income and return on assets; and *cash flow,* including cash throw-off and cash requirements.

How the history of past earnings is figured can make a considerable difference in estimating future earnings. For instance, the owner can make a record of the previous five years and extrapolate a simple average of these (as a buyer might do for negotiating purposes).

Financial Data ($000)

	1966	1967	1968	1969	1970
Sales	$860	$1,550	$1,690	$2,420	$2,940
Net Profit After Taxes	$ 70	$ 145	$ 95	$ 190	$ 225

The average net profit is $145. But this does not give adequate consideration to the growth indicated in the last two years' fine results. Accordingly, the owner should compute a *weighted* average (a buyer might "neglect" to do this), with a weight of 1-2-3-4-5 for each year from 1966 to 1970. The result of this weighted averaging is:

	Actual	Weighted
1966 (weighted 1)	70	70
1967 (weighted 2)	145	290
1968 (weighted 3)	95	285
1969 (weighted 4)	190	760
1970 (weighted 5)	225	1,125
(total weights 15)		2,530
2,530 divided by 15 yields weighted average (earnings potential)		$169

For a company with a good growth record and potential, the weighted average is a truer indication of potential value than is a simple average; and anticipated increases in earnings should be added to the weighted average to give a true expectation of earnings. In addition, *greater earnings potential usually bears a higher price multiple,* particularly if the factors behind the growth are solid rather than non-recurring.

3. *Adjustments for savings.* One of the merger benefits to be sought by both buyer and seller is savings resulting from a good "fit." These savings may derive from reducing personnel costs; more efficient plant operations; better sales and distribution; or other benefits, depending on the objectives of the merger. Just as the buyer estimates the value of these savings, so should the seller. And these savings should be included in his earnings potential, since the buyer will obtain the benefits of these savings after the merger.

A word of caution is suggested, however. In their eagerness to buy, many acquisition-minded companies have experienced disappointment in not obtaining the savings that rosy forecasts had predicted. As a result, they are more conservative in predicting benefits from merging. Estimations that one plus one equals three are now more realistically adjusted to one plus one makes two and a half.

Nevertheless, there must be some mutual advantage or the two firms have no real reason for merging. Accordingly, the seller's potential should reflect an added amount for this value. This gets into the question "of value to whom?" This varies, of course, with each potential buyer.

Price/Earnings Multiple

The price/earnings multiple is the figure paid for the company divided by its after-tax earnings, or the multiple of net profit used to compute the company's purchase price. From a purchaser's point of view, an investor who wants to recoup his investment in ten years would have to realize an after-tax return of 10 per cent (or a pre-tax profit of 20 per cent) on investment, plus adjustment for discounted cash flow and inflation.

Determining an equitable multiple is basically a matter of judgment on the part of the buyer and the seller. Where the seller's risk of doing business is high, the negotiated multiple will be less, and vice versa.

The price/earnings (P/E) ratio of the buyer is important in a sale for stock. These words of caution might be significant to a seller:[5]

> During recent years, many companies, by means of public relations programs, aided by enthusiastic market analysts' reports and public speculative fever, saw their stock soar far beyond prices warranted by prospective earnings

---

[5] George D. McCarthy, *Acquisitions and Mergers.* (New York: The Ronald Press Co., Copyright 1963), p. 79.

and dividend increases. A number of these companies were in the market for mergers and acquisitions. You cannot blame them if they utilized their overpriced stock to buy up acquisition values, which increased their earnings and net assets per share.

On the basis of the foregoing analysis, it may reasonably be concluded that management of a company to be acquired through an exchange of stock by another whose capital stock is selling at a high earnings multiple relative to that of the seller should regard factors other than, or in addition to, market prices in arriving at an exchange ratio.

The owner of a closely-held company can (and should) refer to recent financial journals for average P/E ratios for his industry or industries and for P/E ratios of prospective buyers.[6]

Returning to the preceding situation, where the company has a weighted average of $169,000 in earnings potential, a ballpark valuation of, say, $2.5 million would be fifteen times the weighted average earnings, or eleven times the earnings of the latest year's results. An anticipated growth in earnings would indicate that a higher P/E multiple should be considered.

Another factor in the seller's favor is that since the buyer is usually buying a *going business*—the organization, personnel, market, plant and so forth—and not just the assets, the P/E multiple is based on the ingredients necessary to produce profit. Where there is excess cash beyond that required in the business, such as investments and surrender value of life insurance not required to produce company earnings, these should be pulled out before establishing the P/E multiple. If these had a value of $400,000, the owner in our example should receive this amount plus the negotiated multiple of earnings.[7]

The seller should be aware, however, that the buyer may have to pump in additional financing in order to maintain or improve the company's earnings. The buyer would rightfully lower the multiple accordingly.

Book Value

While book value cannot be disregarded, it is a secondary factor in most valuations. Book value is primarily a means of establishing asset values for accounting purposes. Buyers are interested in book value mainly as a way of determining the general accounting approach, especially as to depreciation, patents, goodwill and expensed

---

[6] See Chapter 6, "Selecting the Right Buyer," for recent price/earnings records of leading conglomerates.

[7] For a concise reference on valuation, see Donald Rappaport, "Buying or Selling a Going Business," in *Handbook of Business Administration,* ed. H. B. Maynard (New York: McGraw-Hill, 1967), pp. 9-113 to 9-127.

items. However, where earnings potential, with appropriate P/E multiplier, comes to less than book value, then book value (adjusted to reflect true value) is obviously the determining consideration in establishing the company's value to buyer and seller.

Adjusted book value takes into account the present net sound value of real estate, machinery and equipment, and an assigned value for items that have been expensed and are not carried on the books.

Appraisal Formulas

Appraisal companies have developed their own formulas for evaluating fair market value. A formula used by one company in processing income into value is this:

Net earnings before taxes (and before return on investment and allowance for depreciation) is computed on a pro forma basis, extrapolated from the previous five years' results. From this are subtracted return on investment (7 per cent on operating capital and 9 per cent on physical assets) and depreciation allowance. The result is "excess earnings attributed to intangible value." This figure is multiplied by a period of years, primarily according to the company's dependence on the top manager (the greater the dependence, the fewer years' multiplier). The value of physical assets is added to this intangible value to give a "fair market value of the enterprise." Current assets, less current liabilities, are added to this figure to give the fair market value of the owner's equity. For example:

Annual pro-forma earnings before taxes		$176,900
Less return on investment:		
Operating capital 7% of $320,000	$22,400	
Physical assets 9% of $208,000	18,720	41,120
Net earnings for depreciation allowance and intangible value		135,780
Depreciation allowance		26,234
Excess earnings attributed to intangible value		109,546
Intangible value, 4 years (rounded)		440,000
Physical assets		208,000
Total physical assets and intangible value		648,000
Current assets less current liabilities		223,000
Fair market value of owner's equity		$881,000

Other Yardsticks

Market price is a measure of value where buyer's and seller's stocks are both traded on national exchanges. Shareholders of the selling company are not likely to agree to a sale at a price less than the indicated market value of their stock.

Potential for paying dividends, based on future results, is a measure considered by many acquirers.

Other valuation factors include comparison of the seller's business with other companies in the same field, value of patents and proprietary products, value of management and technical personnel, and tax-loss carry-over value.

EFFECT OF METHOD OF PAYMENT

The price a seller can get is directly related to terms and the method of payment—cash, securities or a combination. The philosophy of one successful conglomerate reflects the importance of the method of payment: "Give the seller his price, but you (the buyer) dictate the terms."

Because of capital gains taxes, the price the seller will accept, and the price the buyer will pay, are likely to vary considerably depending on whether the sale is for cash or for the buyer's securities. If the owners sell at a price of $3 million, having invested $500,000 in the company, the $2.5 million gain is subject to 25 per cent capital gains tax. The owners would lose $625,000 of the gain at the time of a cash sale.

If the sale is for securities in the buyer's corporation, however, the transaction can be "tax free"—that is, no taxes are due on the capital gain until the securities are subsequently disposed of by the sellers. This is the case where the seller acquires capital voting stock in the acquirer's company. (The tax situation may be different, however, when the seller receives debentures and/or nonvoting stock; in some situations the gain when the securities are sold may be treated as ordinary income and taxed accordingly.) Also, if the sale is for capital stock, a buyer can be much more generous, particularly if his stock is selling at a high price/earnings ratio, since earnings of the acquired company will boost his own earnings base.

In a cash sale the gain could be delayed through an installment payment plan within applicable provisions of the Internal Revenue Code (less than 30 per cent received in the year of the sale, balance spread over a period of years). In this case the buyer should receive a higher price (higher than full cash payment at the time of the sale), since a dollar received later is almost invariably worth less than a dollar received today.

To illustrate the effect of terms on a cash sale, suppose the seller accepts 29 per cent payment at the time of the sale, and a promissory note for the balance spread over five years. One effect is that in the fifth year the buyer will be paying with dollars which, because of inflation, may have about 85 per cent of the purchasing power of the same dollar at the time of the sale. A greater effect than inflation, however, is the present value of a dollar to be received at that time. Only 68 cents, compounded at 8 per cent for five years, is required out of the buyer's pocket at the time of his acquisition in order for him to have the dollar that he will pay five years later. When the effect of inflation is combined with the present value of this dollar to be received five years later, the purchaser is in effect paying with 58-cent dollars. Clearly he can afford a higher price when payment is spread over a period of time.

What the seller will be receiving in exchange for his company, and when he will receive it, are key considerations in the merger transaction. Because of their importance, these matters are discussed in more detail in Chapters 5 and 6.

EFFECT OF ACCOUNTING ASPECTS

The method used by the acquirer to account for the transaction has a bearing on the price the seller will receive. The buyer may treat the matter as a "purchase" or as a "pooling of interests" for accounting purposes.

In a "purchase" (and most purchases for cash, bonds or other nonvoting securities are considered a purchase), the matter of "goodwill"—the amount the buyer pays for assets over their book or appraised value—may not be amortized for tax purposes, and the acquirer therefore may not make deductions against his pre-tax earnings for this item. Also, assets are carried at the current fair market value rather than at the previous carrying value. This requires higher depreciation charges. The effect of these accounting practices is to decrease the purchaser's earnings, as compared to "pooling of interests" accounting.

In a "pooling of interests," usually used in a sale for voting stock, the seller acquires part of the risk of the buyer, and the two organizations are considered for accounting purposes as having always been combined. There is no new basis in accounting for assets, goodwill does not arise, and the buyer's earnings are not reduced on account of the adverse tax effects of these factors. Thus, to a purchaser, "pooling of interests" accounting is usually preferable, and he may therefore be able to offer more (or more accurately, the seller can request more) compared to a deal considered a "purchase." During price negotiations the seller should determine how the buyer plans to account for the transaction.

LIMITS ON THE BUYER'S PRICE

The price a seller will receive is largely a matter of negotiation based on value, terms of payment, effect of taxes and accounting considerations. One other major factor must be considered, however: practical limits on what the buyer can or will pay. The potential seller should be aware of these, since blindness to what a buyer can pay and the reasons for his limitations can close the door very fast—permanently—to what might have been a mutually profitable deal.

A cardinal principle among conglomerates is that in any acquisition there must be "no" dilution in earnings per share. This means that a buyer will evaluate the seller's earnings after taxes, divided by the number of shares he will pay in a stock payment, and compare this with the buyer's earnings per share for the remainder of his business. The buyer will recognize earnings to be achieved in a reasonably short future (potential earnings), and thus the judgment factor enters into his decision. But he has developed sophisticated tools for determining dilution of earnings per share, and he will not tarry long looking at a company whose price would cause dilution, regardless of other attractive features.[8] To indicate the emphasis that experienced acquirers give this point, the presidents of many major acquisition-minded companies refuse to consider *any* dilution in earnings per share. The more liberal professionals take exception to this hard line. They believe that, in exceptional instances, a company could consider dilution by *one or two cents per share*—provided future increased earnings offset this dilution!

Another limitation on price/earnings multiple results from the increased costs that the buyer may incur. For instance, he probably has substantial employee benefits above those provided by the owner of a closely-held company. He will probably have to bring the benefits of the acquisition in line with those of his other facilities, at an added cost to himself. He may include a management contract for the seller's principals in the package offer (although experience shows many owner-tigers "go slack" after selling their interests) and this could prove to be a significant cost. The effect of increased asset base can reduce his expected net income to a point where the price/earnings multiple is out of line for the seller's industry and potential.

Another limiting factor, and a very significant one, is that many buyers aim for getting their investment back within three to five years. This means that both the earnings potential of the merger candidate and the method of financing the acquisition must be very favorable; otherwise, in spite of "fit" and "chemistry," the merger may not measure up to established goals.

And finally, if the proposed acquisition is for cash, the buyer will evaluate very closely whether he will realize a greater return on investment from the proposed acquisition or whether he might do better by putting the money into his existing organization, whose needs and capabilities are much better known to him. If his return on investment is 20 per cent, he is not likely to buy a company that he believes would return less than this.

SUMMARY

Valuation and price are subject to varying measures and to the point of view of the seller and the buyer. The factors discussed in this chapter will enable a seller to evaluate his company's worth on the same bases used by experienced buyers, and to substantiate a reasonable asking price.

But what if the two parties cannot agree on a price? Is the deal off? Not necessarily. There are ways an owner can enhance his company's worth, and there are ways of proving that his company will be as profitable as he says. Both of these are discussed in later chapters. But first, let's take a look inside the buyer's camp and see how he goes about making an acquisition.

---

[8] For example, see Donald J. Smalter and Roderic C. Lancey, "P/E Analysis in Acquisition Strategy," *Harvard Business Review,* November-December 1966, p. 85.

CHAPTER 4

A PEEK INSIDE THE BUYER'S TENT

The merger route is littered with casualities of the feverish activity of this third great merger era. According to various studies by companies directly involved, as well as by consulting firms, investment bankers and other knowledgeable observers, half of the acquisitions completed are disappointing. In a quarter of the total, the acquirer's answer to whether he would acquire a company if he had to do it again is not only no, but hell no. Purchasers' disappointment is confirmed by the rising rate of divestitures.

It can be expected that in a corresponding number of cases the seller is equally disappointed. What goes wrong? How do such rosy expectations on the part of both parties go so haywire?

This chapter outlines what goes on in the buyer's tent in the acquisition process, in order that a seller may understand his potential partner better and in order that his post-merger integration may be less painful than a quarter of those completed in recent history. If it is any consolation to the owner of a closely-held company, the merger process between large companies is at least as awkward as those between a large company and one of his size.

WHY COMPANIES BUY

Reasons for mergers and acquisitions range all the way from whim to survival. Among railroads, the forces of technology, economics and regulations make mergers a practical necessity. An executive who acquired seven companies in a two-year period acknowledges that ego satisfaction in "being up here with the big boys" is his motive. Revenge was the motive in one merger, in which the acquirer bought a company to get at an executive who had sold a large block of the buyer's stock some years before.

Generally, however, mergers are carried out for good business reasons; and it is note-worthy that these have a far greater chance for success.

Diversification is one of the main business reasons for acquisitions. Acquiring additional product lines or marketing channels enables a buyer to level out cyclical swings and to obtain a better balance between industrial or commercial customers and government customers.

Increased earnings, for greater acceptance in financial circles and for greater "acquisition power," is a key motivating factor for many of the widely diversified companies. Increased earnings per share cause the market price of the acquirer's stock to increase; and the increased market price of the stock in turn enables the acquiring company to utilize its stock to buy more earnings. For example, a company whose stock was selling at $22 per share on reported earnings of $1 bought a smaller company for 70,000 shares of common stock. The acquired company had earnings of $100,000 per year or $1.43 per share on the purchase price. The merger increased the earnings per share on the pooled results to $1.03 per share.

Closely related to the effect of increased earnings per share is the added support the financial community may give a stock that shows growth from both acquisitions and operations. Recently the price of some of the leading acquisition-minded corporations ranged from forty to seventy times earnings, for example, while some of the largest industrials who rely on internal growth sold at twelve or thirteen times earnings.

Technology is the rationale for many acquisitions. A buying company recognizes the technical expertise in a smaller firm, realizes that it fits into its own product plans and that it would require too long to develop the capability from within, and makes an offer. The smaller company, pressed for finances and wanting to pursue its technological capability to fulfillment, recognizes that a merger is the way to achieve its goals.

Sometimes the reason for acquiring a company is to obtain its strong management team. A company doing $225 million in sales had a top management team and board whose ages ranged from fifty-nine to seventy-four; average age was sixty-three. Recognizing the need for a strong, young team of replacements, the company paid a handsome price for a company doing $25 million because the acquired company's executives were capable and in their early forties. The acquirer planned to work with these younger executives for two or three years and then move them into the drivers' seats.

Other reasons for buying, each of them reported frequently in business journals, are finding profitable uses for excess cash; gaining a tax loss advantage (the start of Bangor Punta's rising star as a conglomerate); achieving integration (vertical, horizontal, conglomerate) with attendant advantages in utilizing plant capacity, assuring a source of supply and efficiencies in combined operation; acquiring a patent or proprietary position; increasing book value per share; and countering a sudden advantage gained by a competitor.

Whatever the reason given for a merger (and, as with selling companies, the "real" reasons often are not stated), the acquisition process in the better-organized companies then follows a fairly well-defined sequence of events. The question after "why buy" is "how?"

THE ACQUISITION PROCESS

Buying a company is plain hard work. It requires extensive surveillance of possible candidates, intense analysis of the "possibles" culled out, deft handling of contacts and negotiations by the chief executive and a thick skin to absorb disappointment. But these are not enough. No company with extensive acquisition functions (which may go under such labels as "Corporate Development," "Corporate Diversification" or "Long-Range Planning") has avoided making mistakes in at least some of its acquisitions. They have learned much from recent experience, however, and can apply this experience in subsequent transactions, whereas the seller has but one selling opportunity.

In briefest essence, the steps in acquiring a company are these:[1]

Determining objectives to be achieved by acquisition. Having determined what his broad purposes are in buying rather than developing from within (usually these purposes are to save time and to obtain technological capabilities), the buyer develops clearly defined merger objectives and clarifies what he does *not* want.

Setting criteria governing candidates to be sought. Examples of crisp, clear criteria are shown in Figures 4–1 and 4–2. The changing nature of business, incidentally, is illustrated by Bekins' new look. Bekins Van & Storage Co., which is now a subsidiary, became the Bekins Company, a holding company organized in 1968 to diversify into security, maintenance, temporary help, building management and other activities.

Preliminary screening involves developing a list of probably 100 merger candidates in order to obtain one actual acquisition. This list is developed from trade association sources, Dun & Bradstreet reporters, bankers, Poor's, Moody's and other possible sources, including the company's own executives, engineers, marketeers and purchasing personnel. Of those on the list, 90 per cent can be eliminated based on criteria and other factors.

Preliminary investigation is a time-consuming analysis of the remaining candidates, for the purpose of determining one or several with whom to negotiate. This investigation includes analyses of product brochures, financial reports and government and trade publications; field surveys for the purpose of interviewing customers, competitors, bankers, trade associations and others acquainted with the candidate; and a comparison of the relative strengths and weaknesses of each candidate in

[1] For a detailed description, see J. H. Hennessy, Jr., *Acquiring and Merging Businesses* (Englewood Cliffs, N.J.: Prentice-Hall, 1966).

Figure 4-1. Bekins Acquisition (Diversification) Criteria

1. Type of business: Transportation or service-oriented
2. Size of potential company: Sales range $2–20 million
3. Historical growth past five years:
     Industry: Minimum 10 per cent per year in sales
     Company: Minimum 15 per cent per year in sales
4. Profitability:
     Return on sales (before tax) of 12 per cent or greater
     Return on total assets (before taxes) of 20 per cent or greater
5. Capitalization
     Long-term debt-to-equity ratio to be less than 25 per cent
6. Management:
     Must be capable, well seasoned, with proven results
7. Location:
     Headquarters preferably in California

Figure 4-2. Acquisition Objectives

Required Criteria

1. Profitable (rule of thumb: minimum 5 per cent after taxes)
2. In growth markets (sales and earnings projected at 15 per cent annual increase)
3. Potential rate of return of 15 per cent on invested capital
4. Type of business: Manufacturing, distribution or service—excluding strictly consumer-type companies
5. Size: At least $3 million annual sales, unless other circumstances exist, such as a demonstrated rapid growth, advantageous effect on the company or unusually good fit with an existing product or division
6. Favorable method of acquisition (stock, debentures, etc.)
7. P/E multiple of acquired company not more than 75 per cent of parent company.
8. Good management willing to remain or a type of business capable of assimilation into parent company—and managed by parent company's personnel.
9. Earnings per share of the resulting combination should not dilute parent company's earnings per share.
10. Company should have proprietary products as distinguished from "job-shop" operation.

Desired Criteria

11. Preferably a unique product or process
12. Proven marketing and innovative ability in new product development and introduction
13. Preferably in a field with lower competitive factor
14. Product lines should be of a repetitive nature rather than highly engineered, one-of-a-kind
15. In a geographic area conducive to convenient access and management, and one that provides a favorable labor supply

marketing, finance, engineering and development, and manufacturing. A format of the type used in the preliminary investigation is shown in Figure 4–3.

Negotiation opens the direct contact with the merger candidate. It includes initial contact, plant visits, exchange of information and bargaining.

The pre-closing investigation is a thorough audit, study and analysis of the anatomy of the merger candidate to confirm tentative conclusions or to detect discrepancies not previously revealed. It is noteworthy that probably most mistakes in buying or selling a company result not from errors of judgment but from failure to investigate thoroughly. The closely-held company, in particular, can expect to be much more closely scrutinized in the future.

Many companies have developed detailed checklists to avoid being burned. But Litton, one of the most successful acquirers, does not use checklists per se, in the belief that a checklist might become a crutch—and an unreliable one; it depends instead on experienced specialists whose range of investigation is not limited to a checklist (although these experts may have informal checklists of their own). It is worth noting also that a list does not in itself indicate intangibles, such as peoples' attitude toward the selling company, and that these important factors must be sensed and interpreted by an experienced acquisition team.

Because a checklist is used by many buyers, however, and because it may be useful to a seller in preparing for a possible merger, an example is appended to this chapter as Figure 4–4 (pages 51–53). Many checklists are considerably more detailed.

Closing is the meeting in which all documents consummating the merger are exchanged. These include notes, deeds of trust, minute books, stock certificates, audited statements, bylaws, certificates of incorporation, legal opinions, inventories, title policies and, most important, the acquirer's check in payment.

Post-merger integration, as the name implies, consists of dovetailing the organization, functional practices and administrative practices of each company with the other.

The executive time involved in winnowing out the final candidate for merger is enormous. One estimate is 2,600 hours: 4 hours each for determining the initial 100 prospects, 20 hours each for preliminary investigation of twenty of these, 200 hours each for three or four remaining prospects and 1,000 hours in final investigation and closing.[2] Each of these steps requires a higher level of talent, leading to direct involvement of the acquirer's chief executive himself.

A Midwest firm with a particularly good batting average has active files on over 10,000 companies; has made in-depth studies of 850 of these; negotiated with 150; and acquired 37, over a ten-year period. Only 3 of the 37 subsequently became divestitures.

---

[2] *Mergers and Acquisitions, the Journal of Corporate Venture,* Spring, 1967, p. 70.

Figure 4–3. Format for Preliminary Investigation
Profile

1. Names
   A. Corporate*
   B. Trade names

2. Principal location
   A. Other locations

3. Background information
   A. History
   B. Corporate relatives

4. Position in the industry
   A. Describe their position in the industry
   B. What are the trends in the industry?

5. Management
   A. Board of Directors
   B. Officers and key employees
   C. Background, reputation and experience of these people

6. Ownership
   A. Capital structure

7. Marketing
   A. Basic plan
   B. Product lines (services)
   C. Prices
   D. Channels
   E. Market
   F. Market place
   G. Marketing area
   H. Sales organization

8. Manufacturing-Purchasing
   A. Source of their product line

9. Financial information
   A. Sales volume
   B. Profit (losses)
   C. Ratios

10. Why would we want to acquire them?
   A. How would they fit our criteria?
   B. How could they be rearranged to fit our criteria?

Figure 4–3. Format for Preliminary Investigation
Profile—Continued

11. Reputation of the company and their products to these people
   - A.   Employees
   - B.   Suppliers
   - C.   Shareholders
   - D.   Customers
   - E.   Creditors
   - F.   Competitors
   - G.   Distributors
   - H.   Debtors
   - I.   Investment bankers and security analysts

12. What do the indices disclose?
   - A.   Wall Street Journal Index
   - B.   New York Times Index
   - C.   F&S Index of Corporations and Industries (Funk & Scott, Cleveland, Ohio. See also their Predicasts)
   - D.   Business Periodicals Index
   - E.   Applied Science & Technology Index (H. W. Wilson Co.)
   - F.   Electronic News Index

13 What are the apparent leverages that have caused them to succeed or to fail?

14. What are their apparent problems?
   - A.   Internal
   - B.   External

15. What are their apparent opportunities?
   - A.   Internal
   - B.   External

16. What indicates they might be available?
   - A.   Age of owners, shortage of working capital, internal or external problems?

17. Whom should we contact? Why?

18. Suggested approach. Why?

*Corporate names can be very important in both sales and public financing. For example, a company that makes safety devices for scaffold workers changed its name from Safety Tower Ladder Co., to Air Space Devices. See "Growing Companies Play the Name Game," *Business Week,* Feb. 26, 1966, p. 91.

(Reprinted through courtesy of *Mergers and Acquisitions, the Journal of Corporate Venture,* from Robert W. Hollman, "Evaluation of the Corporate Complex," January-February 1968, pp. 60–80.

PITFALLS

The steps outlined above are the ones followed in most organized searches for a merger partner. They sound logical and straightforward; so where do acquisitions go wrong?

At any step along the way.

For one thing, many companies are not organized or prepared for this kind of systematic approach, and wind up buying the wrong company for the wrong reasons. The result is unhappiness for both parties. A veteran of more than thirty acquisitions cites ten key rules for a successful merger, starting with pinpointing objectives:[3]

The Ten "Commandments"

"Must factors"

1. Pinpoint and spell out the merger objectives.
2. Specify substantial gains for the stockholders of both companies.
3. Be able to convince yourself that the acquired company's management is—or else can be made—competent.
4. Certify the existence of important dovetailing resources—but do not expect perfection.

Other key considerations

5. Spark the merger program with the chief executive's involvement.
6. Clearly define the business you are in (e.g., bicycles or transportation).
7. Take a depth sounding of strengths, weaknesses and other key performance factors—the target acquisition company's and your own.
8. Create a climate of mutual trust by anticipating problems and discussing them early with the other company.
9. Don't let caveman advances jeopardize the courtship.
10. Most important of these latter six rules, make people your number one consideration in structuring your assimilation plan.

Fuzzy objectives or nonbusiness objectives are the reason for most acquisition failures, experienced acquirers agree. Unless a company knows where it wants to go, and knows what its own strengths and weaknesses are, how can it hope to find acquisitions that will help it get there? Nevertheless, many try, as the record of sell-offs attests.

As indicated above, a major cause of disappointment is inadequate investigation. Having provided the basic information required by a potential buyer, a seller has often in the past replied to requests for more information, "Why? Don't you trust us? If we don't trust each other at this stage, how can we expect to make an effective merger?" Having received their lumps from bowing to sellers in the past, buyers are

---

[3] Willard F. Rockwell, Jr., Chairman of North American Rockwell, "How to Acquire a Company," *Harvard Business Review,* September-October 1968, pp. 121–132.

becoming more inclined to answer, "Hell no!" when the matter of their trust is questioned. Although experienced acquirers probe for a seller's "real" reason for selling, and try to make a thorough investigation prior to the agreement, it often is not possible to do so because of security reasons. In these instances, buyers will justifiably insert in the terms of agreement that their offer is conditional upon the seller's warranty that his representation is correct, subject to audit. Unfortunately, in spite of this precaution, acquirers sometimes find that they have bought a lawsuit rather than a productive merger partner.

A third cause for disappointment is in "fit." With so many major considerations involved, key "minor" ones are sometimes overlooked, such as whether the selling chief executive will be able to keep his company-paid car, what to do about the seller's pension program, whom the acquired company's chief executive will report to in the combined organization. Enlightened acquirers—enlightened through planning, investigation and the lessons of experience—tend to bend in the direction of keeping the new member of the family happy and productive, even though his practices may not be fully in line with the rest of the corporation.

Another pitfall in the road to merger is the personalities involved. An acquirer may be too aggressive, a candidate too eager or too coy; a key individual on either side may be a pain. In one case, where the acquiring company's legal counsel seemed determined to thwart a promising merger through excessive nit-picking, the head of the selling company called off the deal with, "I'll be damned if I'll sell to a company that would hire a jerk like that!" More often, however, it is the seller's attorney or other specialists who, because of relatively little experience in mergers, are inclined to dwell on small technicalities and lose sight of the purpose of the transaction.

In family-owned companies, delays due to sheer negligence are a notable reason for calling off merger discussions. An acquisition-minded company allows reasonable time for carrying out its plans, but it expects reasonable responsiveness to its overtures.

Sometimes a desirable merger candidate is just too small to see. He is not listed in source materials from which the preliminary list of candidates is prepared and is unknown to individuals contacted; therefore he is never approached by the companies he might like to consider. An alternative open to the small seller, however, is to seek out a desirable partner.

Timing causes suspension of some merger plans. When money is available, the closely-held company is less inclined to merge; when money is tight, a buyer is less inclined to buy, and he will offer less to a seller during a dip in the economy.

One of the major reasons for failure of mergers is that sometimes the management of the acquired company "goes casual" after the merger. Before the sale of his company, the owner's entire estate is wrapped up in the company, and all his business activity reflects directly on this interest. But once the company is merged or sold he is no longer dependent on the success of the company for his personal estate, and as general manager he may take the attitude, "Why should I bust my back when it really doesn't mean that much to me?"

Another cause of failure is lack of follow-through; the acquiring company may not be prepared or staffed to integrate the company into its operations.

FINANCING THE ACQUISITION

How an acquirer finances the purchase of a company is largely his business, and of no concern to the seller—except for the fact that he may well finance it out of the seller's resources. This may not be meaningful to a seller who receives full value for his company, and who has no further operational connection with it. But because he may have an earn-out arrangement, or because he may want to continue in an operational capacity after his company becomes a subsidiary, the seller should know something about the techniques of financing a purchase.[4]

In a bootstrap operation the assets of the acquired business are used to the extent possible as security for repaying the debt incurred by reason of the purchase price. This burden often is assessed against the acquired company; and the former owner, now manager, may find that his parent company will postpone normal expenditures for maintenance, capital equipment and R&D costs that will not bear fruit during the payback period.[5] A qualified broker will probe the buyer's intent to prevent this kind of situation.

Another form of bootstrap, which may be desirable to a seller because of tax reasons, is to receive installment payments of the selling price. He can accept up to 30 per cent in the first year after sale, and spread the remainder over a number of years. The seller might also receive a good rate of interest on the unpaid installments (although, as noted in the next chapter regarding discounted cash flow, he should evaluate the value of today's dollar against a dollar to be received in the future).

A third form of bootstrap, mentioned in Chapter 3, might occur where, unknown to the seller, he is in effect giving away his business and is being paid for part of the value of his assets.

Other than these sources from the seller, a buyer has basically two other choices in a cash acquisition: He can provide equity capital directly from his own resources or from a stock offering, or he can arrange debt capital from third parties. The latter is often done when the acquirer himself is a relatively small company. It is significant that lenders are more interested in the acquisition's potential ability to repay the debt than they are in the net worth shown on the balance sheet; this is the same viewpoint as acquirers, as noted in the previous chapter.

---

[4] See Donald Rappaport, "Buying or Selling a Going Business," *Handbook of Business Administration,* ed. H. B. Maynard (New York: McGraw-Hill, 1967), pp. 9-120 to 9-121.

[5] Robert L. Chambers, "How Not to Sell Your Company," *Harvard Business Review,* May-June 1961, pp. 105-108.

Where the acquisition is for securities, the buyer has a fairly broad range of financing techniques. The seller must evaluate as carefully as he can the real worth of what he is getting as an investment.

THE MERGER-MINDED COMPANIES

It is reported that 70 per cent of mergers involve widely diversified companies (who now generally avoid the term "conglomerate"). In one year alone (1968) the most active acquirers completed these merger transactions:

Acquirer	No. of Mergers
U.S. Industries, Inc.	28
Teledyne, Inc.	20
Instrument Systems Corp.	20
Republic Corp.	19
Gulf & Western Industries	17
Whittaker Corp.	15
Crowell-Collier	15
Fuqua Industries	14
Walter Kidde & Co.	12
Anixter Brothers	12
Transitron Electronic	11

In a two-month period, U.S. Industries accomplished six mergers; Walter Kidde & Co., Alco Standard Corp., Georgia-Pacific Corp., Mid-Continent Telephone Corp. and Omega Equities Corp. finalized four mergers each; six other corporations completed three mergers each; and twenty corporations completed two mergers apiece.

MERGER TRENDS

As might be expected in a period when so many mergers are consummated and so many acquirers with varying degrees of experience are active, divestitures are a particularly active part of the merger game.[6] Acquisitions that proved to be disappointing are being sold in whole or in part (for instance, product lines), partly due to anticipated government action requiring profit reporting by divisions. Some companies are reshaping their corporate structure and image, and in the process are selling or spinning-off parts that do not fit. Therefore the "rebound" market can be expected to become increasingly active.

It can be anticipated that Federal regulatory bodies will take additional action to slow the merger activity of large companies who might conceivably have a

---

[6] "The Corporate Sell-Off," *Mergers and Acquisitions, the Journal of Corporate Venture,* November-December 1968, p. 58.

restraining effect on trade; this may lower glamour stocks' prices and increase the number of shares expected by a seller. But smaller companies who have no particular business reason to engage in acquisitions, as well as those who do have good reasons, will probably more than make up for any decline in larger companies' merger pace. This means that there will be more medium-size fish in the pond, who will have a keen eye for the minnows and who in turn will be attractive morsels for the big fish.

Capable, flexible management will be a highly sought commodity as shakeouts and resignations take a toll among merged organizations. At the same time, corporate transplants will be accomplished more deftly, with more of a scalpel and less of a hatchet, so that there will be fewer dislocations among key individuals.

And finally, the merger fever can be expected to continue as long as the economy continues to grow. With simultaneous explosions in technology, population and international trade, the economic outlook for the remainder of the twentieth century looks bright indeed.

Columnist Art Buchwald describes how a conglomerate is born in "The Making of a Conglomerate—a Buchwald Fantasy."[7]

Dalinsky's Drug Store in Georgetown decides to merge with Fischetti's Meat Market in Bethesda. Dalinsky and Fischetti can't agree on which name to use, so they call the company The Great American Drug and Meat Company. A stock offering is immediately sold out.

They take over the Aetna Curtain Company, Markay Life Insurance Company, Mary Smith Pie and Bakery Company, Winston Life Preserver Company, Washington Green Sox Baseball Club, the Norfolk (basketball) Warriors, and a TV station in rapid succession, because, "if you stand still, you die." Then came a bank, another bank, a mutual fund, a fried chicken franchise company, and so forth. In less than three years their original $55 investment has provided $50 million apiece on paper, and control of $3 billion worth of businesses.

Buchwald concludes, "The only danger is that if either Dalinsky Drug Store or Fischetti's Meat Market loses the lease on its store, the whole conglomerate pyramid could fall down. When you get right down to it, that's the only part of their business that Dalinsky and Fischetti understand."

---

[7] The Washington Post Co., copyright 1968.

Figure 4–4. Diversification Check List

Note: The following is a checklist developed by the Rockwell Manufacturing Company, Pittsburgh, Pa., to be used as a guide and reminder when investigating companies for possible purchase. Rockwell cautions that neither this nor any other checklist can replace the sound judgement that comes from experience, but that it has proved practical and valuable as a working tool and as a reminder to check points that might sometimes be overlooked in an investigation.

A. *General*

 1. Statement of proposed transaction and objectives
 2. History of business and general description
 3. List of officers and directors; affiliation
 4. Stock distribution—number, principal holders, etc.
 5. Organization chart
 6. Policy manual

B. *Financial*

 1. Latest audited financial statements
 2. Last available financial statements
 3. Ten-year summary financial statements
 4. Projected operating and financial statements
 5. Full description of securities, indebtedness, investments and other assets and liabilities other than normal day-to-day accounts
 6. Chart of accounts and/or description of accounting practices relative to inventories, fixed assets, etc.
 7. List of bank accounts, average balances
 8. Credit reports from banks and Dun & Bradstreet
 9. Federal income tax status; i.e., excess-profits-tax credit, any loss or unused EPT credit carry-forwards, latest year audited, any deficiency claims, etc.
10. Summary of state and local tax situation; i.e., applicable taxes, unemployment tax rate, any deficiency claims, etc.
11. Tax status of proposed transaction; recommendation for best method of acquisition
12. Complete list of insurance policies, including description of coverage and cost; workmen's compensation rate
13. Statement of responsible officer of business as to unrecorded or contingent liabilities
14. Nature of inventory

C. *Sales*

 1. A brief description and history (if any) of the product line
 2. A ten-year record of product sales performance
 3. A long-range forecast of growth or contraction trends for the industry of which the product line is a part.
 4. A three- to five-year forecast of anticipated demand for the product
 5. An estimate of the industry's ability to supply present and anticipated demand
 6. A three- to five-year forecast of sales expectations for this company (share of the market)
 7. An analysis of the effect of anticipated increased volume and/or cost reduction on:
   a. Product demand and share of the market
   b. Market saturation and overcapacity
 8. An analysis of the effect of the geographic location of the new facility on:
   a. Product demand and share of the market

Figure 4-4. Diversification Check List—Continued

    b. Distribution costs (freight savings, warehousing, etc.)

    c. Competitive position

9. A review of present sales management, selling force, advertising and sales promotion policies for adaptability and adequacy in relation to new facility

10. A review of present competitors and competitive practices including:

    a. Description of *competitive* products

    b. Location

    c. Estimated share of market

    d. Pricing policies

    e. Methods of distribution

11. An analysis of present and/or probable pricing policies for the product line considering:

    a. Competitive position

    b. Cost pricing

12. An analysis of present and potential domestic and export customers:

    a. Major types of customers and per cent of sales to each

    b. Geographical location

    c. Buying habits

D. *Manufacturing*

1. Description and layout of plant property
2. List of principal machine tools—age and condition
3. Opinion re maintenance and "housekeeping"
4. Utilities—availability, usage, rates
5. Estimated total annual fixed cost
6. Organization, departmentalization
7. Transportation facilities
8. Description of area, including climate, hazards from flood, etc.
9. Opinion re adequacy of auxiliary equipment—tools, patterns, material handling equipment, etc.
10. Detailed expense schedule
11. Building codes, zoning laws, and restrictions

E. *Purchasing*

1. Principal materials used
2. Relation of material costs to sales
3. Purchasing methods
4. List of principal suppliers, items, location
5. Inbound freight costs
6. Workload—last twelve months:

    a. Number of purchase orders issued

    b. Value—purchase orders issued

    c. Value of outstanding commitments

F. *Research and Engineering*

1. Description and condition of facilities:

    a. Drafting room and office

    b. Experimental room

    c. Laboratory

    d. Special test equipment

2. Engineering personnel—quality and quantity of technical talents. . .employed. . .unemployed
3. Product designs—evaluation; condition of drawings
4. Patents and trademarks—coverage, existing applications, litigation

Figure 4–4. Diversification Check List—Continued

G. *Labor*
 1. Number, sex and age—present employees
 2. Direct, indirect, administrative; number and cost
 3. Number of potential job applicants from surveys or census
 4. Determination of types of skills available in the area from state employment service and other sources
 5. Location and availability of students from high schools and technical schools
 6. Union—copy of contract
 7. Labor relations history
 8. Appraisal of working conditions
 9. Statistics on turnover; reasons
10. Description of incentive system; average rates incentive and hourly
11. Employment and personnel policies
12. Accident frequency
13. Ratio of total labor cost to sales
14. Pension and welfare plans
15. Appraisal of transportation, community recreation facilities, housing
16. Evaluation of labor situation in area

CHAPTER 5

EARN-OUT

After the total "fit" is established, the stumbling block to most merger deals, naturally enough, is price; a seller values his company at $4 million, and a buyer values it at $3 million. As discussed in Chapter 3, the terms of a merger may overcome part of the difference and a bending by each party may help close the gap. But if a half-million dollar difference remains, how can the parties effect an agreement? Is this the end of the "fit" that promised to provide such mutual benefit to each of the parties?

In many cases involving closely-held companies this has, in fact, been the end of the courtship. There is a way, however, to bridge the gap between the sparkling potential claimed by the owners and the more pessimistic one as seen by the buyers: the earn-out. The earn-out is a way of rewarding the sellers, particularly in closely-held companies, based on earnings subsequent to the merger. As sellers become more familiar with its possibilities, the earn-out is expected to become a more frequently used way of bridging this gap between offer and acceptance.[1]

One effect of an earn-out is that it makes for realism in both seller's and buyer's negotiating positions, and consequently enables both parties to realize opportunities that might otherwise be lost. However, unless the seller knows its ramifications and knows how to measure the value of a future dollar versus one received today, the buyer may have an advantage in its use. Setting the future earnings goal is the moment of truth for the seller; if he sets his goal too high, and falls short, he will receive something less than maximum price for his company. If he sets it too low, the buyer will offer less during negotiations.

---

[1] See Charles J. Hecht, "Earn-Outs" in *Mergers and Acquisitions, the Journal of Corporate Venture,* Summer, 1967, pp. 2–12.

HOW THE EARN-OUT WORKS

In an earn-out the purchaser must be willing to leave profit and loss responsibility with the seller, since achieving the merger's profit aims is the basis for the earn-out payment. This has an additional benefit (in most cases) to both parties: The new acquisition will be run with a minimum of the disruptions frequently inherent in mergers. (This can be a disadvantage if the formerly tight-ship owner-operators become lax. This is unlikely, however, in view of the earn-out carrot.) Another condition in an earn-out is that the newly acquired business must be distinct (for example, a division or autonomous subsidiary) from the parent company. And where the earn-out is to be paid in stock of the parent (the usual case), the sellers must obviously have (well-founded) faith that the buyer's star will continue to rise; this hammers home the necessity for the seller to view the merger transaction as an investment.[2]

Given these conditions, let's see how an earn-out works, with an assumed merger of Standard Industries into the buyer, Buyrite Corporation. Standard, a manufacturer of automobile parts and owned principally by three members of the Stannard family (two sons and a daughter of the founder), has shown a notable increase in earnings since introducing a line of sports car accessories several years earlier, and is considered by its owners to be worth $4 million in a merger. Buyrite, a rising conglomerate, wants to diversify its enterprises into the automotive field in view of the potential of the industry and Standard's size and earnings compared to competitors. However, based on analysis of the outlook for the industry, audit of Standard's market, books, production, and so on, Buyrite arrives at an offering price of only $3 million. Negotiations bring the seller's asking price down to $3.7 million, and the offer up to $3.2 million.

At this point the two parties discuss a five-year earn-out arrangement. This is the first direct acquaintance the Stannards have had with this type arrangement, and they are naturally watchful as to what the experienced representatives of Buyrite will suggest. The two parties have generally agreed on a sale of Standard for common stock in Buyrite; but the principles of earn-out would be equally applicable in a cash sale or in a combination cash and stock sale. The Buyrite negotiator suggests a stock settlement of $2.5 million on completion of the merger, and an earn-out that could provide up to 60,000 additional shares. At the current price of $25 per share, this would be an additional $1.5 million.

Earn-Out Formulas: The Base Period Earn-Out

In this formula, the Stannards would receive additional stock for each year of the five-year earn-out period in which earnings exceed a previous base period (such as the

---

[2] For a look at selling a company as an investment, see Richard M. Hexter, "How to Sell Your Company," *Harvard Business Review,* September-October 1968, pp. 71–77.

previous high-earnings year, or the year immediately preceding the merger; or a simple or weighted average of years, if profits fluctuate significantly). In the 1968 base year agreed upon as indicative of future growth, the company earned $300,000 after taxes. The earn-out for this type formula is

$$\frac{(\text{Excess Earnings}) \times (\text{Capitalization Rate})}{\text{Market Price of Buyrite Stock}} = \begin{array}{c}\text{Additional Shares} \\ \text{for Stannards}\end{array}$$

where excess earnings is Standard earnings above the $300,000 base; and the capitalization rate is computed as the initial down payment ($2.5 million) divided by Standard's after-tax earnings for the base period, $300,000, or a capitalization rate of approximately 8. Note the significance of a high initial payment on future earnings: If the Stannards had accepted an initial down payment of only $1.8 million, the capitalization rate would be only 6. The market price is Buyrite's price as of the end of each earn-out year; the higher the price the fewer shares of stock for the Stannards, and vice versa.

Assume earnings for Standard prove to be as follows for the five years after selling, and Buyrite stock remains at $25 per share:

	After-Tax Earnings	Excess Over $300,000 Base	Then the Stannards would receive
1969	$300,000	$0	— 0 shares
1970	$350,000	$50,000	$\dfrac{\$50,000 \times 8}{25} = 16,000$ shares
1971	$400,000	$100,000	$\dfrac{\$100,000 \times 8}{25} = 32,000$ shares
1972	$325,000	$ 25,000	$\dfrac{\$25,000 \times 8}{25} = 8,000$ shares
1973	$350,000	$ 50,000	$\dfrac{\$50,000 \times 8}{25} = 16,000$ shares
			72,000 total shares

The Buyrite negotiator may slip in a limit on shares per year, however, equivalent to one-fifth of the 60,000 total additional maximum shares proposed, or 12,000 maximum shares per year. In this event the Stannards would receive only 44,000 shares, since they would receive none in 1969; maximum of 12,000 shares in 1970; 12,000 in 1971; 3,000 in 1972; and 12,000 in 1973.

The number of shares will also vary according to the price of Buyrite stock; this can be a very favorable factor if the stock continues to rise, as follows:

	After-Tax Earnings	Excess Earnings	Buyrite stock increase $5 per share each year
1969	$300,000	$ 0	$30 per share 0 shares
1970	$350,000	$ 50,000	$\dfrac{\$50,000 \times 8}{35} = 11,450$ shares
1971	$400,000	$100,000	$\dfrac{\$100,000 \times 8}{40} = 20,000$ shares
1972	$325,000	$ 25,000	$\dfrac{\$ 25,000 \times 8}{45} = 4,450$ shares
1973	$350,000	$ 50,000	$\dfrac{\$50,000 \times 8}{50} = 8,000$ shares

With the same maximum of 12,000 shares per year, the Stannards would receive no shares in 1969; 11,450 in 1970; 12,000 in 1971; 4,450 in 1972; and 8,000 in 1973—a total of 35,900 shares. *The market value of these shares, however, would be considerably higher at the 1973 stock price than the greater number of shares at a steady $25 per share price, because Buyrite's price increase more than offsets the fewer number of shares received* (35,9000 shares at $50 versus 44,000 shares at $25). This is an important consideration to the Stannards, who may later want to sell or exchange some of their Buyrite stock in order to diversify their capital. Knowing the future for one's buyer in an earn-out is obviously an important bit of information for the seller—and is usually never questioned in most mergers.

There is also built-in protection to the seller, to a degree, if the stock drops; in this case he will receive more shares of the buyer's stock, up to the maximum per year agreed upon. If Buyrite stock dropped to $20 for each year, the Stannards would receive 46,000 shares of stock versus 44,000 at the $25 price.

The Increment Earn-Out Formula

A type of earn-out that is becoming more popular, particularly with acquiring companies intent on increasing earnings per share from each acquisition, is a formula whereby additional shares are earned by the seller for each year he exceeds the previous high year's earnings. Using the preceding data for the base period earn-out, the Stannards increased their after-tax earnings in the 1969–1970 annual increment ($50,000); increased again in the 1970–1971 increment ($50,000); declined the following year; and increased from 1972 to 1973, but not up to earnings previously reached ($400,000 in 1971). Therefore they would earn (at $25 per share):

$$1970 \quad \frac{50,000 \times 8}{25} \quad \text{or} \quad 16,000 \text{ shares}$$

$$1971 \quad \frac{50,000 \times 8}{25} \quad or \quad 16,000 \text{ shares}$$

or a total of only 32,000 additional shares (24,000 shares, if the maximum of 12,000 per year is applied). This type formula obviously puts more of a burden on the seller, who is betting that (1) he can forecast each year's earnings trend, (2) the general economy will not adversely affect him and (3) the price per share he receives will be worth the heavy responsibility he continues to shoulder. For most sellers, this particular formula clearly is a doubtful attraction.

The Cumulative Earn-Out Formula

In a cumulative earn-out, the cumulative earnings in excess of base period earnings is the basis for additional shares of stock. In the example, where base-year earnings are $300,000, the cumulative earnings expected would be $1.5 million for the five-year period. The Stannards can receive additional stock, depending on how earnings range between this $1.5 million expected minimum and a pre-agreed maximum, say $2 million. In this case excess earnings for the five-year period total $225,000. They would thus receive 225,000 divided by the difference between 2 million and 1.5 million, multiplied by 60,000 (maximum possible), or 27,000 additional shares of stock.

$$\frac{225,000}{2,000,000 - 1,500,000} \times 60,000 = 27,000$$

The "pre-agreed maximum" in a cumulative earn-out is the key negotiation point; it will vary considerably from one industry to another, according to optimistic but achievable earnings possibilities.

In this kind of earn-out the Stannards would be "locked-in" for the five-year period, with disadvantages to the seller of no additional stock until the end of this period, and disadvantages to the buyer of having to live with Stannard management, which may during this time become less effective because of removal of incentives and because of personal reasons (health, desire to retire, age).

The Profit-Unit Formula

In this earn-out, additional Buyrite stock is earned on the basis of each increment of earnings (for instance, each $10) over the earnings base. In the Stannards' case, this would be

		Additional Shares
1969		0
1970	50,000 ÷ 10	5,000
1971	100,000 ÷ 10	10,000
1972	25,000 ÷ 10	2,500
1973	50,000 ÷ 10	5,000
		22,500

The key figure in this case is the "increment of earnings"; the sellers should compute the effect of various increments before agreeing or disagreeing with application of this concept.

The "Annual Average" Profit-Unit Formula

A modification of the previous earn-out provides for additional stock to be earned according to net profit increments in excess of a specific average for three to five years. For example, the formula in one merger provided for additional stock of the acquirer up to 50,000 shares where the average annual net before taxes exceeded $400,000 over a three-year period.

Additional shares earned are based on incremental earnings of $10,000, and based on the price of the purchaser's stock as averaged for several trading days prior to the anniversary date of the closing, as follows:

Average Earnings for Three Years Before Federal Income Taxes	Additional Stock (Maximum 50,000 Contingent Shares)
$400,000–$410,000	$ 30,000
410,001– 420,000	90,000
420,001– 430,000	150,000
430,001– 440,000	210,000
770,001– 780,000	2,250,000
780,001– 790,000	2,310,000
790,001– 800,000	2,370,000

The "Marko" or Reverse Earn-Out

Harold Marko, President of Soss Manufacturing Company, described his "Minimum-Risk Merger Method" to a group of acquisition-minded conferees in a 1967 seminar in New York.[3] The gist of the scheme is that if a company sold to Soss realizes less than a previously set earnings average, the seller will forfeit part of the purchase price, which is set aside in escrow for this purpose.

EFFECT OF SALARY ON EARN-OUT

Where the seller continues in a management position after a merger involving earn-out, a reduced salary may be more than offset by his earning additional stock. In one case the owner had received a $50,000 annual salary, and expenses for operating his plane (amounting to $10,000 annually) were charged to the company. After the merger his salary was reduced to $35,000 and he paid for plane expenses out of his income. These changes made a pre-tax difference of $25,000 additional profit for the

---

[3] Edward T. O'Toole, "Marko's Minimum-Risk Merger Method," *Mergers and Acquisitions, the Journal of Corporate Venture,* Winter, 1967, pp.55–57.

acquiring company. If his earn-out provided for two or three dollars or more in additional stock for each additional dollar of pre-tax earnings, his additional stock would be a multiple of this $25,000 savings—a strong incentive to accept a lower current income.

DISCOUNTED CASH FLOW

Closely related to earn-out is the question of what a dollar to be received later—up to ten years later—is worth today. Clearly, even aside from the effects of continuing inflation, a dollar in the hand is worth more than a dollar to be received later. But how much more?

Figure 5-1 shows the present value of a dollar received at various times in the future, at various rates of discount. For example, the first figure under the 8 per cent discount column is .926, which means that the present value of $1 to be received a year from now is 92.6 cents. In other words, using a rate of 8 per cent, $0.926 on hand today is equivalent to $1 a year from today; $0.857 (next figure in the 8 per cent column) on hand today is equivalent to $1 two years from today (assuming in each case that there is absolutely no risk); $0.794 (next figure in the column) on hand today is equivalent to $1 three years from now; and so on. This is the same as saying that $0.681, invested at 8 percent compound interest, will be worth $1 five years from today; or, in a merger, the present cost to the purchaser is only 68 cents for each dollar to be paid five years from now.

The basis of discounted cash flow is the earning power of the dollar today. If the sold business earns 20 per cent on invested capital, and the seller accepts payment (in whole or in part) five years later, the acquirer in effect is making a present payment of 40 cents for each dollar to be paid later.

Let's see now how discounted cash flow works in practice, and what its effect is on the sellers of a closely-held company.

Using the Marko Method, the President of Soss Manufacturing Company acquired the Mueller Steam Specialty Company, a manufacturer of metal strainers and valves, in 1964, with a five-year earn-out.[4] The maximum price to be paid was set at $2.25 million and the minimum set at $1.25 million, with a sliding scale in between depending on earnings. The buyer estimated that he could have bought the company for $1.85 million if a settlement at the time of sale could have been worked out.

Because of the former Mueller owners' diligence and the benefits of "fit," it appears that they did, in fact, receive the $2.25 million maximum price in 1969, or $400,000 more than the apparent 1964 price. Referring to discounted cash flow to see how much the $1.85 million in 1964 would have been worth in 1969 versus the $2.25 million maximum, it can be determined that for each dollar received in 1969 the owners could apparently have received $1.85 ÷ $2.25 in 1964, or 82.2 cents. Refer now to Figure 5-1. Looking at the five years hence line, and .822, it is apparent that the

---

[4] *Ibid.*

Figure 5–1. Present Value of $1

Years Hence	1.0%	3.0%	4.0%	5.0%	6.0%	8.0%	10.0%	12.5%	15.0%	17.5%	20.0%	25.0%	30.0%	35.0%	40.0%	50.0%
1	.990	.971	.962	.952	.943	.926	.909	.889	.870	.851	.833	.800	.769	.741	.714	.667
2	.980	.943	.925	.907	.890	.857	.826	.790	.756	.724	.694	.640	.592	.510	.510	.444
3	.971	.915	.889	.864	.840	.794	.751	.702	.658	.616	.579	.512	.455	.406	.364	.296
4	.961	.888	.855	.823	.792	.735	.683	.624	.572	.525	.482	.410	.350	.301	.260	.197
5	.951	.863	.822	.784	.747	.681	.621	.555	.497	.446	.402	.328	.269	.223	.186	.132
6	.942	.837	.790	.746	.705	.630	.564	.493	.432	.380	.335	.262	.207	.165	.133	.088
7	.933	.813	.760	.711	.665	.584	.513	.438	.376	.323	.279	.210	.159	.122	.095	.059
8	.923	.789	.731	.677	.627	.540	.467	.390	.327	.275	.233	.168	.123	.091	.068	.039
9	.914	.766	.703	.645	.592	.500	.424	.346	.284	.234	.194	.134	.094	.067	.048	.026
10	.905	.744	.676	.614	.558	.463	.386	.308	.247	.199	.162	.107	.073	.050	.035	.017
11	.896	.722	.650	.585	.527	.429	.351	.274	.215	.170	.135	.086	.056	.037	.025	.012
12	.887	.701	.625	.557	.497	.397	.319	.243	.187	.144	.112	.069	.043	.027	.018	.008
13	.879	.681	.601	.530	.469	.368	.290	.216	.163	.123	.093	.055	.033	.020	.013	.005
14	.870	.661	.577	.505	.442	.340	.263	.192	.141	.105	.078	.044	.025	.015	.009	.003
15	.861	.642	.555	.481	.417	.315	.239	.170	.123	.089	.065	.035	.020	.011	.006	.002
16	.853	.623	.534	.458	.394	.292	.218	.152	.107	.076	.054	.028	.015	.008	.005	.002
17	.844	.605	.513	.436	.371	.270	.198	.135	.093	.064	.045	.023	.012	.006	.003	.001
18	.836	.587	.494	.416	.350	.250	.180	.120	.081	.055	.038	.018	.009	.005	.002	.001
19	.828	.570	.475	.396	.331	.232	.164	.107	.070	.047	.031	.014	.007	.003	.002	
20	.820	.554	.456	.377	.312	.215	.149	.095	.061	.040	.026	.012	.005	.002	.001	
21	.811	.538	.439	.359	.294	.199	.135	.084	.053	.034	.022	.009	.004	.002	.001	
22	.803	.522	.422	.342	.278	.184	.123	.075	.046	.029	.018	.007	.003	.001	.001	
23	.795	.507	.406	.326	.262	.170	.112	.067	.040	.025	.015	.006	.002	.001		
24	.788	.492	.390	.310	.247	.158	.102	.059	.035	.021	.013	.005	.002	.001		
25	.780	.478	.375	.295	.233	.146	.092	.053	.030	.018	.010	.004	.001	.001		
26	.772	.464	.361	.281	.220	.135	.084	.047	.026	.015	.009	.003	.001			
27	.764	.450	.347	.268	.207	.125	.076	.042	.023	.013	.007	.002	.001			
28	.757	.437	.333	.255	.196	.116	.069	.037	.020	.011	.006	.002	.001			
29	.749	.424	.321	.243	.185	.107	.063	.033	.017	.009	.005	.002				
30	.742	.412	.308	.231	.174	.099	.057	.029	.015	.008	.004	.001				
31	.735	.400	.296	.220	.164	.092	.052	.026	.013	.007	.004	.001				
32	.727	.388	.285	.210	.155	.085	.047	.023	.011	.006	.003	.001				
33	.720	.377	.274	.200	.146	.079	.043	.021	.010	.005	.002	.001				
34	.713	.366	.264	.190	.138	.073	.039	.018	.009	.004	.002	.001				
35	.706	.355	.253	.181	.130	.068	.036	.016	.008	.004	.002					
36	.699	.345	.244	.173	.123	.063	.032	.014	.007	.003	.001					
37	.692	.335	.234	.164	.116	.058	.029	.013	.006	.003	.001					
38	.685	.325	.225	.157	.109	.054	.027	.011	.005	.002	.001					
39	.678	.316	.217	.149	.103	.050	.024	.010	.004	.002	.001					
40	.672	.307	.208	.142	.097	.046	.022	.009	.004	.002	.001					
41	.665	.298	.200	.135	.092	.043	.020	.008	.003	.001	.001					
42	.658	.289	.193	.129	.087	.039	.018	.007	.003	.001						
43	.652	.281	.185	.123	.082	.037	.017	.006	.002	.001						
44	.645	.272	.178	.117	.077	.034	.015	.006	.002	.001						
45	.639	.264	.171	.111	.073	.031	.014	.005	.002	.001						
46	.633	.257	.165	.106	.069	.029	.012	.004	.002	.001						
47	.626	.249	.158	.101	.065	.027	.011	.004	.001	.001						
48	.620	.242	.152	.096	.061	.025	.010	.004	.001							
49	.614	.235	.146	.092	.058	.023	.009	.003	.001							
50	.608	.228	.141	.087	.054	.021	.009	.003	.001							

Muellers received the equivalent of *4 per cent* compound interest! (This can be verified by taking a slide rule and moving the slide and hairline alternately, multiplying by 1.04 five times.) This is doing only slightly better than keeping pace with inflation.

In actuality, the owners received $1 million at the time of the sale and a split in profits (which Soss retained as return on investment), with a balloon payment at the end of the earn-out period; and this conceivably could have raised the owner's return to a compound interest of 7 1/2 per cent. But this is more than offset by their risk—a possible $600,000 loss if they had earned only the minimum price—and by their having borne five years' profit responsibility at a time when they would probably rather have gone fishing or traveling.

Thus it appears that the Mueller owners could have invested their money better, without the operational responsibilites and risk and with no reverse earn-out liability, by taking the $1.85 million in cash or in cash and securities.

Moral: Probe the value of a future dollar versus a dollar received today. Even without considering inflation or the "enjoyment-now" factor, the seller can take a substantial loss when his compounded interest rate is compared with what he could receive on other investments.

DEFINITION OF EARNINGS

What constitutes earnings in an acquired company is vitally inportant in an earn-out formula. Where earnings are computed "in accordance with generally accepted accounting principles," we will see (Figure 7–3, "Accounting Magic," pages 92–93) what can happen within this meaning; and where there are operational changes, as seems inevitable in almost all mergers, the definition of earnings can make a great difference in earn-out and can lead to friction between the buyer and former owners.

Briefly, some of the points requiring definition are:

Taxes—whether pre-tax or after-tax earnings are considered; and who will receive previous years' tax liabilities and credits discovered after the merger.

Depreciation—whether the new subsidiary is to be saddled with an increased depreciation rate, reducing its earnings during the earn-out period.

Salaries—assurance by the seller that he will not change salaries and other benefits unless approved by the buyer; and in fact that he will not initiate any major personnel actions that would cause loss of efficiency.

Corporate charges—acquiring corporations will often assess a charge, such as one per cent of the subsidiary's gross, for management services provided. Where the acquiring corporation provides capital to the acquired firm, the corporation will often charge the base rate of interest plus one per cent to the acquired company.

CERTIFICATES OF CONTINGENT INTEREST

The earn-out is peculiarly applicable to closely-held companies because of the relative ease in getting agreement among the owners. In publicly-held companies, however, it is possible to effect the same incentive between the buyer and seller by use of the certificate of contingent interest (CCI). A detailed description of how a CCI works is given in the November-December 1968 *Mergers and Acquisitions* by Charles J. Hecht; and an example of a CCI, used in the merger of Hayes International Corporation into City Investing Company, with accompanying notes by the same author, is given in the Summer, 1967, issue of the same publication.

CONCLUSION

The earn-out is likely to increase in use as sellers become more aware of its potential and its risks. In dealing with a buyer it would be well to look at the form of earn-out and each factor used in the formula, with trial applications to determine the effects of changing each factor. And where it appears that an earn-out is favorable and is a logical way to close the price gap, it is also advisable to determine the effect of a lesser cash or stock amount now versus a dollar five years from now. Five years may be later than you think ... and the earning power of your "dollar" meanwhile may be amazingly low.

CHAPTER **6**

SELECTING
THE RIGHT BUYER

"The vast merger literature that has accumulated over the years has been curiously silent on the interests of sellers. Advice on planning, objectives, strategy and negotiation is usually expressed exclusively in buyers' terms. . . . "[1]

"In no other area of managerial judgment is there so much potential for disaster. Selling companies frequently underprice themselves, or lose out by ignoring some important tax or financing wrinkle. . . ."[2]

"There seems to be no company—which either has sold or is currently available for sale—with a "Department of Divesture" or a "Director of Being Acquired." Buyer after buyer, however, has established the opposite number. . . ."[3]

These implications for the seller are underlined by innumerable instances where the seller sold out to the wrong buyer, or where it was perhaps a "right" buyer but the wrong deal, or frequently to the wrong buyer and for the wrong deal.

How can a seller determine who among prospective merger partners is "right"? How can he approach the right one? What should he evaluate about a potential merger candidate? How and when should he make company information available? How can he evaluate whether a right merger partner is proposing a "right" deal? These are difficult questions, particularly in view of the discreet circumstances that must ordinarily surround the proposed sale of a closely-held company, and in view

[1] Warren G. Wintrub (editor), *Planning Business Combinations,* Lybrand, Ross Bros. & Montgomery, 1968, p. 12.

[2] "Learning the Ways of Matchmaking," *Business Week,* April 13, 1968, p. 127.

[3] Richard M. Hexter, "How to Sell Your Company," *Harvard Business Review,* September-October 1968, p. 71.

of the limited experience of most chief executives in this crucial business decision.

This chapter will assist the owner of a closely-held company to select the right buyer. Inherently this means not only choosing the right company, but also evaluating his offer—knowing how to value his stock if (as in most cases) he offers stock in exchange, and, if the seller is concerned with the post-merger "fit" of the two companies, some policy and organizational aspects to consider.

SELECTING A BUYER: THE APPROACH

A selling company will usually command a higher price if it meets the special needs of an acquirer than it will if it meets only a general need, such as good earnings potential or general diversification. Therefore, a seller should determine early in the conversation with a buyer, "Why do you want us? What do you think we could do for you?" The answers will indicate what points the seller should stress in negotiations and in preparing his company for earning the best price; and the answers will also show whether the acquirer has done his planning homework and where the seller fits on the desirability scale.

Determining whose special needs he meets brings up a problem: Should the seller wait until, hopefully, the two or three most suitable companies approach him? Is there an in-between route? Or should he wait and maybe accept an offer based on some company's general need? The plot of the courtship is thick indeed, with knowledgeable advocates of each approach.

The Sadie Hawkins Approach

In "How to Sell Your Company" the author makes a case, backed by successful experience, for aggressively searching out a merger mate.[4] He says that by waiting for approaches by buyers the seller has lost initiative, the advantages of timing, choice of financial arrangements and a possibly broader selection of partners.

The procedure in such an approach is this: Characteristics of prospective buyers are determined—complementary markets, products, ability to provide financial support, growth potential, an organization that provides a good fit and an ability to pay the seller's price. Then, probably with the help of a broker or other outside specialist, the seller screens an initial list of fifty or more companies in order to select a handful of most likely prospects. Each of these is then analyzed in more detail, in much the same way that a buyer searches for a seller. Comparative analyses are made of the remaining contenders, and finally a "winner" is selected. He is then presented with the opportunity to buy, with reasons why he should and why he was selected.

In his experience, the writer states, the list prepared in this way has never failed to produce a merger. In an example, where the Diogenes Corporation (seller) wants to achieve in three years a market value of $50 million in the equity of an acquirer,

---

[4] *Ibid.*

three contenders are analyzed (Figure 6–1) and a "winner" selected, in this case Honesty, because of the lower initial price and the greater latitude in negotiations available to Diogenes' merger team. Honesty's president is then approached as the prospective buyer.

Figure 6–1. Comparative Analysis of Three Possible Buyers of Diogenes Corporation

A. Ability of proposed buyers to meet certain criteria

	Quality image	Business fit	Corporate personality fit	Ability to pay	Investment attraction
Beauty Corporation	Favorable	Very favorable	Favorable	Very favorable	Neutral
Honesty Corporation	Favorable	Very favorable	Favorable	Favorable	Very favorable
Wisdom Corporation	Very favorable	Favorable	Neutral	Favorable	Favorable

B. Investment attraction of proposed buyers

	Compound annual growth		Earnings per share		Price/earnings multiple		Stock price		
	1962-1967	1967-1970*	1967	1970*	1967	1970*	Current	1971*	Appreciation potential
Beauty Corporation	26%	15%	$2.40	$3.65	30	24	$71	$87	22%
Honesty Corporation	14%	13%	$1.25	$1.80	21	22	$26	$39	50%
Wisdom Corporation	5%	17%	$2.50	$4.00	22	19	$55	$76	38%

C. Current price required to have $50 million market value in 1971

	Market value 1971 (in millions)	Appreciation potential	Market value 1968 (in millions)
Beauty Corporation	$50.0	22%	$41.0
Honesty Corporation	$50.0	50%	$33.1
Wisdom Corporation	$50.0	38%	$36.2

*Estimated

The Find Me Approach

The Sadie Hawkins approach is immediately suspect, however. In the same publication the chairman of North American Rockwell says, ". . . no company is ever for sale, at least not admittedly. . . . A company on the block is like a girl in search of a husband. If she asks the fellow to marry her, she frightens him off. If she's too cold and aloof, he looks for a warmer climate. There is a delicate balance to be maintained."[5]

This seems to be the general consensus among most experienced acquirers. Being approached by a sell-belle has some of the aspects of this refrain:

> I love you. Therefore I am a lover.
> All the world loves a lover.
> You're all the world to me.
> Therefore you love me.

[5] Willard F. Rockwell, Jr., "How to Acquire a Company," *Harvard Business Review,* September-October, 1968, p. 121.

[or, in mergerese]

I want somebody to buy us.
Everyone's buying.
You're the greatest.
Therefore let's merge.

Most companies, considering the times a seller's market, wait to be approached by an acquirer. It becomes a game among some, who compare notes on how many times they are approached during the week, without seriously contemplating merger.

But one problem in playing a waiting game, for the company that is at least somewhat interested in a merger prospect, is that the company may not be "visible" to desirable merger mates. Its size alone may make it hard to find; or perhaps the fact that it would complement the strengths of a potential acquirer has not yet been perceived by him.

Another problem in waiting is that timing is a key factor in mergers. In anticipation of regulatory action governing some aspects of mergers (for example, FTC action regarding conglomerates), some companies are stepping up the tempo of mergers now, because they may be slowed-down later;[6] and "later" may find a seller, playing the waiting game, still waiting. Another aspect of timing is that the rate of mergers varies with the economy, in both its major and minor fluctuations. For instance, the present merger wave parallels the good economic period of 1947 to the present; while 1958, a recession year, was one in which sellers fared less favorably than in better years. In "How Not to Sell Your Business" Robert L. Chambers says, "Don't wait too long! Mañana may never come. When your company's future is sunniest may be the best time to sell."[7]

Nevertheless, most would-be sellers are reluctant to take either the Sadie Hawkins or Find Me approach because they require more time and sophistication than may be available, and particularly because this may drive down a buyer's offering price.

The Indirect Approach

Between the shy and the brazen approaches is the indirect approach. For many closely-held companies this is the best way to seek out a buyer, without being overtly involved in the process. The approach is through an intermediary, such as a merger packager or other discreet, reliable third party. One big advantage is that this individual can approach a buyer and induce him to come to the party, without revealing the selling company's name or allowing it to be shopped around. It is important

[7] Robert L. Chambers, "How Not to Sell Your Company," *Harvard Business Review,* May-June 1961, pp. 105-108.

[6] An article in the *Wall Street Journal,* September 1967, carries this heading: "*The Merger Surge:* Firms Join at Fast Clip, Spurred Partly by Fear U.S. May Blow Whistle."

that the medium selected be one that will produce buyer possibilities without the seller's intent reaching the ears of his employees, competitors, suppliers or customers.

If he is careful, and knows his contact very well, an owner could plant the seed himself. For instance, he might mention to a service club friend, "You know, if Consolidated had our products, and we had their marketing organization, we could both make a bundle."

What Information to Release? When?

With the exception of a brief tour through the seller's facilities, most contact with a buyer-candidate is usually done off the seller's premises. This means that limited information is provided until the buyer makes a full investigation (often after the final merger agreement is signed); and until the buyer signs a letter of intent, this limited information is as much as he should be given. After all, he may be dealing at the same time with a competitor or the seller's banker. A profile of what this limited initial information consists of is shown on page 26. Within this general context of information the seller can give details and answer questions freely; but an audit or detailed investigation is out of line at this point. This procedure enables the chief executive to control the transaction; once the acquirer's accountants and other specialists move into the merger procedure, the chief executive has effectively lost control over who gets information and how much.

Limiting information to this extent appears to be contradictory to the trend among acquiring companies that *more* information is required, in view of failures of mergers due to insufficient pre-merger investigation. In reality, however, most experienced buyers at this point are in the process of making a broad survey of merger partners, without wanting to invest the time in an in-depth investigation. As Rockwell says, "The suitors who most often wind up as bachelors are the ones who carry on the most elaborate studies. By the time they identify the ideal mate, someone else has grabbed her first!"[8]

Discussion of price should *follow* a discussion of each party's goals and the advantages of merger to each. This sequence has benefits for both parties: If the objectives of a merger fail to develop, there is no point in discussing price, and possible disclosure of price information to unauthorized individuals is avoided. On the other hand, if it develops that both parties stand to gain from the merger, having explored the advantages will tend to keep the merger discussions moving toward this goal, even when price subsequently becomes a problem.

KNOW YOUR BUYER AND HIS PROPOSAL

If a seller contemplates a lump-sum cash deal, he can get his money, pay taxes on it, and that's that. He does not need to know anything in particular about the

[8] Rockwell, *op. cit.*

buyer or his plans for the future. This is what the owner of a small electronics company did in the early 1950's, getting a million dollars cash. But this seller could have done somewhat better if he had been gifted with foresight or had known the potential of his buyers. Today, if he had taken stock instead of cash, his investment would be worth $150 million or more. The seller's name: Charles Litton.

A banker cites a similar recent instance in which the seller of a company doing $8 million in sales received $4 million in cash. The buyer, whose sales were $10 million, had a million dollars in debt prior to the transaction, and $5 million contingent loan liability afterward; this precluded him from raising finances for subsequent acquisitions, and he floated a stock issue instead. Does this sound as though the buyer was inept, at least as to financing? Perhaps, but this is a separate issue. The point here is that the seller in this case also neglected to look into the buyer's potential. Within seven months after the merger the buyer's stock went from 16 to 70, with indications of a continued rise.

On the other hand, a seller may be receiving "Chinese money" when he exchanges his company for the stock of a diversified, fast-growing, acquisition-minded company. Recognizing this possibility, some merger prospects refuse to consider an offer for stock "unless the buyer is on the New York Stock Exchange . . . or at least the American Stock Exchange!"

Securities

This section outlines the main factors to consider in determining a buyer from the standpoint of his offer; and because most offers are for holdings in the acquiring company, it provides an overview of some of the most active multimarket companies (who reportedly account for 70 per cent of the merger activity), their track records, and an expert's opinion on the merits of being listed or nonlisted. Clearly, very few sellers in the past have analyzed the future of their investment. It is assumed, however, that unless the chief executive's mind is made up— "cash only, don't bother me with facts"—he will, as a matter of good business and in his responsibility for making the best possible business investment, consider the merits of accepting securities in exchange for the sale of his company.[9]

Changing Times, the Kiplinger Magazine, for April 1968, contains an excellent summary of conglomerates as an investment:[10]

A conglomerate might prove a sound investment if you can manage to sort

---

[9] "Unless the sale is related to ultimate income and inheritance tax effect before its completion, the owner runs the risk of irremediable losses through the payment of unnecessary taxes." *Planning Business Combinations, op. cit.,* p. 15.

[10] "Conglomerates: The Merger-Minded Wonders of Wall Street," excerpted by permission from *Changing Times,* the Kiplinger Magazine, April 1968. Copyright 1968 by The Kiplinger Washington Editors, Inc., 1729 H St. NW, Washington, D. C. 20006.

out companies with solid growth potential from the flash-in-the-pan enterprises. To be sure, you could say the same of chemical companies or electric companies. But, unfortunately, sizing up a conglomerate is harder because a number of special factors have to be taken into consideration:

How much "dilution" in earnings? Typically, these companies finance part of their mergers by issuing securities that are convertible into common stock. Should the holders of those securities decide to convert, the company's earnings would have to be shared among more stockholders. Earnings per share would automatically be less. One key figure to look for in the company's reports, then, is what professionals call the "diluted" earnings—what earnings per share would be if all convertible securities were turned into common stock. In some companies, full conversion, plus the exercise of stock options given to officers, would reduce earnings per share by as much as 20 per cent to 30 per cent.

How much does each activity contribute? A mere listing of a company's products and services can be highly misleading unless you know how much each of them contributes to the corporation's sales and profits. Some products have larger profit margins and more promising markets than others. Examine the breakdowns of sales and earnings by product lines or company divisions; they are often included in annual reports and in the analyses prepared by advisory services, such as Standard & Poor's and Moody's.

Where is the company going? Does it have a long-range development plan or is it just buying companies helter-skelter? What fields does it want to enter? You should be able to derive some idea of the management's strategy, if there is one, from the annual reports.

How much of the growth is internal? A lot of a conglomerate's growth is bound to come from acquisitions. But no matter how shrewdly the management may wheel and deal in buying companies, the ultimate test of its skill is its ability to increase the profitability of the bought companies after they join the fold. Brokerage house reports, particularly those prepared for institutional investors, such as banks, sometimes divide sales figures to show the "internal" gains, the progress made by previously acquired companies, and the "external" gains, the income added during the year simply by the purchase of new companies.

Who's in charge? Most investors have only a passing knowledge of the people running the companies in which they own shares. But you will notice that many of the reports on the conglomerates devote a good deal of attention to the background and outlook of the officers. The reason is that your investment in a conglomerate is often tied to the coattails of a few top men. Make sure, therefore, that you look into their records, too, before you buy stock in their companies.

Figure 6–2 shows the type of financial data an owner considering merger might evaluate in determining whose offer seems most attractive.

Figure 6-2. The Acquisition-Minded

Potential Earnings Dilution*	Company	Sales (Millions) Latest 12 Months	1962	Per-Share Earnings 1968†	Est. 1967	1962	March 1969†	Price Jan. 1968	Dec. 31 1962
12.6%	"Automatic" Sprinkler	$ 161	$ 19	$0.10	$2.20	$0.04	20	67	7¾
30.3	Avnet	147	31	1.24	2.33²	0.77	24⁶	60	19
39.6	Bangor Punta	147	14	2.55	3.00³	1.91	39	60	9¾
23.1	Boise Cascade	504	175	2.38	1.85	0.68	66	47	11¼
1.9	Borden	1,589	1,048	1.66	2.20	1.53	32	34	29
0.6	Consolidated Foods	1,090	520	1.86	2.65²	1.21	42	56	23
6.6	Continental Telephone	146	9	1.08	1.20	0.27	24	29	10
7.9	Eltra	404	240	2.80	2.70	1.01	39	37	10½
3.8	General Telephone	2,391	1,328	2.10	2.25	1.14	38	46	23
4.5	General Tire	990	960	2.49	1.85	1.55	29	28	21
8.4	Georgia-Pacific	675	325	3.30	2.65	1.34	91	60	26
9.2	W.R. Grace	1,420	547	3.00	2.75	1.89	40	43	39
4.0	Greyhound	578	360	1.35	1.60	0.96	22	22	15½
23.0	Gulf & Western	645	66	3.54	3.91²	0.40	37	64	7¾
3.3⁵	ITT	2,180	1,090	2.58	4.55	2.41	52⁷	105	42
4.7	Leaseway Transportation	177	37	1.95	1.50	0.74	32	24	11
22.7	Litton Industries	1,560	394	2.08	2.60²	1.31	56	101	29
18.9	MSL Industries	112	46	1.99	3.00	1.97⁴	38	34	22
9.0	Raytheon	709	581	2.10	3.75	2.24	40⁷	95	27
3.6	Singer	1,083	631	4.53	4.75	3.28	76	63	63
19.4	Teledyne	392	10	1.61	2.75	0.12	43⁸	136	11½
6.2	Textron	1,194	550	2.10	2.00	0.74	37	50	7½
3.8	United Utilities	145	56	1.25	1.25	0.80	29	29	15

*Potential dilution in earnings per common share if all convertible bonds and preferred were converted, and if all stock options and warrants were exercised (based on latest available date). ¹Initial offerings price on Nov. 18, 1965. ²Actual. ³On fully taxed basis. ⁴Before tax adjustments. ⁵Excluding options

*Potential dilution in earnings per common share if all convertible bonds and preferreds were converted, and if all stock options and warrants were exercised (based on latest available data). †Added by authors. ¹Initial offering price on Nov. 18, 1965. ²Actual. ³On fully taxed basis. ⁴Before tax adjustments. ⁵Excluding options. ⁶After 3-for-2 split in 1968. ⁷After 2-for-1 split in 1968. ⁸After 2-for-1 split in 1969.
Source: "The Executive Investor," prepared by Moody's Investors Service for *Dun's Review*, February 1969, p. 101. Reproduced by permission.

Conglomerates are speculative, to varying degrees. This is partly because their image as being in a particular industry or technology can change so radically, and partly because a lot of the interest in them results from speculation that there will be a quick rise in price as compared to demonstrated capacity for long-term profits.

It is generally agreed that depth in management is a key factor in market appraisal of conglomerates. Noel Hemming, conglomerate specialist of Harris, Upham & Co., rates as strong-management companies International Telephone, Textron, TRW, Ling-Temco-Vought, Lear Siegler and Teledyne, "just about in that order of preference."[11] Other smaller, well-managed conglomerates include Doric, Zero Manufacturing Co., Royal Industries, Maxad, and Health Tecna.

> Ultimately, it is by meshing skill in operations with skill in acquisitions that allows a company to maintain its high price/earnings ratio and in this way convert paper value to real assets. In other words, a part of annual earnings increases can be attributed to using "Chinese money" (stock with a relatively high P/E) to buy companies with lower price/earnings ratios. However, unless a firm can improve operations in acquired companies, it will soon find its growth limited, because (a) the size of acquisition required to boost earnings will become ever larger; and (b) the acquiring company will find it increasingly difficult to find willing sellers if it has a reputation for doing nothing with, or even milking, companies it buys.[12]

In the area of general evaluation of his proposed buyer, the seller has some advantages over the average investor. For instance, the seller is in a position to ask that the proposed buyer provide balance sheet, profit-and-loss statement, history of price/earnings trends and earnings-per-share information (this information is immediately available if the company is publicly held); and he can inquire as to the company's plans for the future, profit contributions by major segments of the company (although legislation is contemplated that would make this information available to the public) and where his company would fit in the picture. In particular, he can *check with other companies whose owners previously sold out to the same buyer.* From them he can inquire how favorable a deal they received; how the stock of the acquiring company has fared since then; whether the acquirer lived up to promises made and whether the mutual advantages expected have actually materialized; whether the previously acquired company has become more or less profitable; and, if they had it to do over again, whether they would sell to the same buyer.

---

[11] Noel Phillip Hemming, quoted in Joseph d'Aleo's "The Conglomerates: A Wall Street Reappraisal," *Mergers and Acquisitions, the Journal of Corporate Venture,* May-June 1968, p. 49.

[12] George H. Zwerdling, Management Consultant with the Boston Consulting Group, "In Defense of Conglomerates," *Mergers and Acquisitions, the Journal of Corporate Venture,* May-June 1968, p. 56.

Usually the acquirer's chief executive will be dealing with the selling chief executive; the latter should take advantage of this rare opportunity to raise well–thought-out questions regarding the future of his investment. One such question might relate to anticipated effects of Federal regulatory action on the market value of stock that would be received in a sale for stock. One observer notes that the price of some glamour stocks are based on earnings projections not for just ten to twenty years into the future, but into the hereafter. It is difficult to see how such optimism would fail to be tempered by regulatory action. Another question might be who the acquirer's major competition is, or will be. A third question might pertain to the acquirer's plans for developing operational strength in his corporation.

A NOTE ON STOCK EXCHANGES

A professional experienced in mergers of closely-held companies frequently encounters this conditional statement: "I don't care what else [about the buyer's deal], but he must be listed on the Big Board!" This reflects an understandable desire on the part of the seller to get something solid for the solid value he is giving in exchange. But to knowledgeable investors, including the individual noted above, it reflects a naïveté about stock markets, in that:

1. A New York Stock Exchange (NYSE) listing does not ensure that the stock received will grow, that it is invulnerable to the effects of competition, that it is not speculative or that its management will be as astute as possible, decade after decade. In fact, all of the major companies listed on the major exchanges today achieved their most important growth *before* their shares were listed. Of companies listed today on the NYSE, only a few remain of those listed half a century ago; some perished or were themselves acquired in mergers. "But a sufficient number fell by the wayside to demonstrate that stock listing is no miraculous key to immortality or even to corporate health."[13]

2. The American Stock Exchange (Amex) generally lists younger companies than does the NYSE; yet it "lists many of the finest and oldest companies in America. It is not, as some have imagined, a list of speculative growth stocks. Annually, 65 per cent of the companies on the Amex pay cash dividends and about a quarter have paid dividends without interruption for from 10 to 100 years."[14]

3. As to over-the-counter (OTC) or unlisted stocks, the article notes that "such blue-chip firms as Anheuser-Busch and Rockwell Manufacturing Company and most of the nation's banks and insurance companies, which are our largest financial

---

[13] Leroy Pope, "To List or Not to List?" *Mergers and Acquisitions, the Journal of Corporate Venture,* January-February 1968, pp. 6–16.

[14] *Ibid.*

institutions, prefer to remain unlisted"; and "OTC dealers quote some rather impressive figures tending to show that their better stocks have outperformed good listed stocks in terms of price and price/earnings ratios in recent years."[15] With regard to whether a company's stock will do better if listed on a national exchange, the article notes that some do, some do not. It notes that some have prospered as a result of listing but that others have not: of forty stocks that moved to the Big Board in 1963, only eighteen were selling at a higher price three years later. This is not in itself conclusive; but for growth and security of investment resulting from sale of his company, an owner should take off any blinders regarding advantages or disadvantages of a particular exchange, and should look to the merits of the stock proffered.

As conglomerates get bigger they become interested in bigger potential acquisitions; meanwhile, junior conglomerates are forming and sprouting like weeds. Information on the latter group may be less easily available from published sources, but some of these may offer particularly good investment opportunities. Again, there is no substitute for knowing the company's record, its plans, capabilities and its key operating personnel. And the best source is likely to be the man who has sold them his firm.

ORGANIZATIONAL FIT

The question of "fit" is important if the selling chief executive or other owners anticipate that they will serve as executives in the merged company. In this case they will need to know, in addition to the soundness of their investment, where the newly acquired subsidiary will fit in the corporate organization; the profit, sales and other objectives expected, and whether the acquisition will have an opportunity to take part in setting these goals; what changes will be expected during the first year's operations; benefits (executive and all personnel) provided, and what disposition will be made of the seller's benefit provisions; to whom the selling chief executive will report; controls and reports that will be instituted; and, particularly, details as to how the benefits of the merger will be achieved. This last might begin to involve functional heads of the selling company with their counterparts in the acquirer's organization.

In "How to Acquire a Company" the author suggests a way to determine the mutual benefits to be attained.[16] The method can apply to selling, as well as to acquiring, companies. He suggests developing a list of "fit factors" for the two potential partners, such as:

[15] *Ibid.*

[16] Rockwell, *op. cit.*

- Provide a much needed diversification opportunity

- Fill a gap in technical or scientific expertise

- Produce financial capabilities for capital expansion

- Strengthen an internal management or operating weakness

- Buy valuable time for management (for instance, to allow sufficient development time for a product-line change)

- Provide a system of reporting that will produce faster and more comprehensive management decisions

- Open new market capabilities

- Reduce the company's dependence on a limited field of growth

- Produce important savings through the ability to make volume purchases

- Shore up a vulnerable patent position

- Increase the stock's price/earnings ratio

- Provide better and more reliable quality control

- Give a needed boost to the company's image and reputation in the marketplace

- Provide important research facilities and expertise

- Allow management to reap the benefits of a tax-loss carryover

- Flesh out a product line for either or both companies

- Put excess management or physical facilities to use

- Provide needed management expertise in marketing, finance, acquisitions, international operations or other fields

He concludes by noting that "the more such values apply to both partners, the greater the chances that the merger will succeed."

GOALS AND PLANNING IN SELECTING a BUYER

The foregoing sections of this chapter skirt the fringe of the key points that are mandatory in any company sale: *It is absolutely necessary that the seller determine his goals and carry them out through execution of a well thought out plan.* Unless his business and personal goals are defined, he will find it extremely difficult to determine whether he should sell for cash or stock; and if for stock, how to evaluate the securities he would receive, and how the securities of one buyer compare with those of another; and whether he wants to remain in an operational capacity with the merged company, and if so what to expect in operating as head of a division or subsidiary.[17]

In setting his objectives, the seller will be doing both himself and prospective purchasers a favor. The latter have been known to remark, almost wistfully, "We'd like to walk in and talk intelligently once in awhile."

CONCLUSION

"It is important for each prospective seller to define his own goals specifically in terms of the amount of consideration sought, the rate of return on capital desired thereafter; the degree of risk acceptable and the extent of management participation."[18] The key to selecting a buyer, then, starts with the seller's determining his own objectives. After this step, selecting a buyer is a matter of planning and of evaluating both the would-be acquirer and his offer.

In the past many sellers have been mere opportunists in making this ultimate investment decision. A former company president says, "I am amazed at how naïvely many presidents go about the job of selling their companies . . . . They approach the job with less thought and finesse than their greenest salesman shows when he makes his first sales call."[19] One major difference: The president has only one opportunity to do a good job in making *his* sale.

---

[17] Warren G. Wintrub (editor), *Planning Business Combinations,* Lybrand, Ross Bros. & Montgomery, 1968, p. 12.[66] For the most effective implementation of his decision, he needs the most explicit possible definition of his true objective in order to develop the most effective selling strategy. *In a surprisingly high number of cases, would-be sellers fail to make their true objective sufficiently explicit to identify all the alternatives and thus insure optimum results.* And the incidence of failure—though differently measured—may well equal or even exceed that for acquirers." (Italics added.)

[18] *Ibid,* p. 14.

[19] Chambers, *op. cit.*

CHAPTER 7

INCREASING YOUR COMPANY'S VALUE

An acquiring company's veteran of more than twenty purchases described what he would do if he were an owner preparing his company for a merger.[1] If he had three years to prepare his company for sale, he would establish an increase-in-earnings trend as follows: In the first year he would spend more than normally on research, plant and machine maintenance and long-term promotions, to keep reported profits down. Conservative accounting practices, such as expensing items that would ordinarily be capitalized, would help lower profits from a normal $200,000 for the year to $100,000. In the second year the company would operate in its usual way with less expensing, in order to bring profits up to $200,000. In the third year he would reverse the accounting practices of the first year and would make every possible effort to reduce costs—development costs, maintenance, promotion expenses—in order to show a $300,000 profit.

Negotiations would then proceed on this earnings trend, with a higher jump-off figure ($300,000) and greater indicated potential. The price/earnings multiplier might be fifteen rather than ten or twelve, and with a higher base if a weighted average is used.

Buyers generally are getting more sophisticated in their investigations, because of being burned in the past; but practices such as this might still be overlooked, in the urge to merge, by less-experienced acquirers.

Basically there are three ways of increasing the value or selling price of a

---

[1] As described in Myles L. Mace and George C. Montgomery, Jr., *Management Problems of Corporate Acquisitions*. (Boston: Graduate School of Business Administration, Harvard University, 1962).

company: By giving it the appearance of added value, as the preceding example indicates; by enhancing its true value; and by the owner knowing what he should get for his company.

APPEARANCE VALUE: MERCHANDISING

A good presentation goes a long way toward making a buyer willing to start negotiations at a relatively high level. It makes sense, therefore, for owners to package their company in an attractive way, the same as they would its products or services.

An executive in the acquisitions group of a leading conglomerate illustrates the value of an attractive appearance. "Sometimes," he says, "we'll look at a company, and the production area is so clean you can eat off the floor. This means that the organization has real leadership, that the people believe in what they're doing and have pride in their work. On the other hand," he continued, "there's the kind of company that two of our people made a cross-country trip to look at. We had a real interest in pursuing the possibility of a merger until our representatives got a thirty-second look at the outfit. In half a minute, viewing operations from a balcony and seeing how dirty, cluttered and poorly managed the area was, they knew this company was not for us."

The moral: An attractive appearance is essential in keeping a potential acquirer interested. A good appearance may not necessarily reflect the true value of the selling company, but a slovenly appearance almost certainly does.

Often a company's management, accustomed to seeing their facilities every day, are so intent on the activities that go on within the walls that they fail to see a poorly groomed exterior, disorderly administrative and operational areas and signs of wear and tear. An outsider, such as a good friend who heads another company, can be helpful in making an "inspection" and in suggesting needed improvements that accustomed eyes would fail to see.

Dressing up a company takes on many aspects, including changing business practices to make profits appear better than they would normally be. Often, in the closely-held company, this is a matter of adopting accounting and other practices of publicly-held companies where professional managers are charged with the responsibility of looking after the stockholders' interests. The profits of the smaller, management-owner company would naturally be greater when similarly recast to reduce the cost of personal benefits charged against the company. For tax purposes the company would continue its present practices, but in preparing for merger it should have a merger broker or CPA recast its statements to reflect a higher profit. Experienced buyers like to work with companies who do this homework prior to the start of negotiations.

Another aspect of getting his company ready is for the owner to review his *real* reasons for selling—the areas that he, better than anyone else, knows are weak—and to improve these facets of the business.

Sometimes, in contrast to the three-year program mentioned earlier, a company's chief executive will embark on a crash program to reflect higher profit within only a few months. He might do this by renting equipment that would normally be purchased, whittling back advertising and promotion expenses while increasing temporary sales inducements and sharply reducing research and development costs. Although these actions would slow a company's growth over a longer time span, the immediate effect is to increase profit and thus the company's selling price. In one instance this kind of action increased after-tax profit by a half-million dollars, resulting in a purchase price of $2 million more than would otherwise have been realized.

On the other hand, in a similar instance where the merging company's chief executive practiced similar accounting procedures, his corporation had such an appearance of robust good health and growth that he became chief executive of the combined corporation. Later, when deceptive accounting and other short-term practices were admitted and the stockholders of the buying company brought court action, an out-of-court settlement was reached that brought the price paid more in line with value received.

Because of recent instances such as this where an actual loss operation has been disguised as a profitable one, most acquiring companies require substantial warranties as to the selling company's conditions and conduct a thorough investigation as soon as time permits. Inventories in particular are likely to receive scrutiny. (When asked what advice he would offer other acquisition-minded companies in avoiding a "buy" mistake, the president of one well-known conglomerate says, "Inventories, inventories, inventories!"—excess, obsolete, overpriced or just plain missing.) Changes in practice to enhance a seller's profit appearance are less likely to go undetected than was the case in the mid-1960's.

A comparison of ratios from one year to another is particularly revealing. For instance, a potential buyer may look at such ratios as these for the previous five to ten years:

- Sales expenses to sales
- R&D expenses to sales
- Sales to inventory
- Maintenance expenses to equipment valuation
- Overhead to sales
- General and administrative expenses to sales
- Inventory turnover (cost of sales divided by average inventory)
- Sales to net worth
- Gross margin

Where there are significant deviations from a trend, particularly in the year or two preceding a proposed sale, the analyst can raise some very pertinent and penetrating questions.

INCREASING THE COMPANY'S BASIC VALUE

Increasing the company's basic value requires one to three years' time, and requires concentration on profit planning and, simultaneously, building on strengths and shoring up weaknesses. Highlighting a few practical principles may be worthwhile to assist the owner preparing for future mergerability. These are probably only a reminder of the management principles he already knows. But many a company founder *knows* what to do to make his company grow beyond its $3 to $5 million size; it is just that, for one reason or another, he does not *practice* these elements of managerial know-how. The situation corresponds to that where the agriculture expert offered advice to a farmer. The farmer replied that he did not need all that advice because "I'm only farming half as good as I know how already."

Profit Planning

A banker says, "Lack of planning, including financial planning, is the biggest lack in closely-held businesses." As recently as a dozen years ago, he could have made the same statement about most of the biggest companies, where profit planning was haphazard and casual. Even today, while planning in functional areas and general objectives has become more formal and effective, profit remains the bottom line, depending on the vagaries of each of the preceding lines to determine whether or not it comes up to expectations. This is just the reverse of true profit planning, where the company's profit goals are stated first; and on subsequent lines the sales objectives and costs of doing business are detailed and shaped to yield the profit goal.

An example of the potential effect of profit planning is the short history of Scientific Data Systems (SDS) of El Segundo, Calif. The company was founded in 1961 by its president, Max Palevsky, and eleven other executives with backgrounds in the computer industry. It lost more than $400,000 on sales of $1 million in its first year. But it reversed this result soon afterward; and by the end of 1968 it showed an after-tax net of $10 million on sales of $100 million. In early 1969 a merger of SDS with Xerox was announced, with an exchange of stock that provided a price/ earnings multiple of ninety-four to holders of SDS stock.

The company's president says, "We don't have many second thoughts about where the action is. It's clearly on the bottom line. The only other criterion for success is how long a term or how short a term. But the bottom line is the key element."[2] Starting with corporate profit objectives, planning extends down

---

[2] Jack B. Weiner, "SDS: It's Real-Time Management," *Dun's Review*, September 1968, p. 58.

through divisions and departments, with profit and sales objectives set at each level, and with each responsible manager's objectives set down in writing. Results are compared regularly with the plan, and action is taken as necessary to ensure that SDS's profit goal is reached.

A key principle followed in profit-minded companies is the setting of "hard" objectives—that is, objectives that are quantified and have a completion time—as distinct from "fuzzy" objectives. A typical example of a fuzzy objective is "increased profits," which is realized whether the company makes one dollar more or ten times as much profit as in a previous year. Numbers communicate, and unless objectives are stated in terms of numbers and have specific dates assigned, the question whether or not the objectives were achieved is a matter of opinion that can vary considerably between owners and lower-level management.

A second principle in planning is that objectives should be challenging rather than routine or easily attainable. This requires the guts to be measured and held accountable for failure or success. The reason management by objectives is resisted in many organizations is that management does not have the courage to put specific goals on record and to take the blame or credit for results. Nevertheless, the growth of profit by design, rather than by chance, requires the setting of challenging objectives in both line and staff elements.

A third principle in planning is that the doers must have a major part in the planning process. A company may have an individual responsible for planning, in the sense that he coordinates individual plans (functional plans) with each other and with the corporate plan; but to be successful, plans must be developed by the individuals responsible for carrying them out. This means pushing objectives and planning down to the lowest possible level, even down to the individual worker's job objectives. In developing this kind of planning, a manager or supervisor should ask, "How does this operation contribute to profit?" If the question is too difficult to answer, the operation may be one that should be eliminated.

Even staff activities and other difficult-to-measure functions can be measured against specific criteria. The objectives for advertising, for instance, can be defined in terms of how many individuals it reaches in the market the company is trying to reach, and the attitude of the individuals toward the company and its products.

A fourth principle of profit planning is to compare actual performance regularly with planned performance, and to take action accordingly, resisting the temptation to lower targets in order to meet lower-than-expected performance.

Building on Strengths; Overcoming Weaknesses

The combination of strengths and weaknesses is unique to each firm. A starting point for increasing the company's basic value is determining where these strengths

Figure 7-1. *How to Analyze Your Own Business*

PROGRESSIVE CONDITION	AVERAGE CONDITION	WEAK CONDITION

1. GENERAL

Policy

PROGRESSIVE CONDITION
1. Company's policies and objectives clearly defined and understood by all.
2. All current and potential economic factors recognized in overall company planning.
3. Aggressive participation in trade and business associations.
4. Company kept currently informed on Federal, State and local regulations.

AVERAGE CONDITION
1. Company's general policies not clearly defined nor understood.
2. Sporadic consideration of potential economic factors in company planning.
3. Interest in trade and business associations limited.
4. Government relationships determined by legal counsel but not passed along to company executives.

WEAK CONDITION
1. No general policy except to carry on company's tradition.
2. Planning done under impulse.
3. Trade and business associations regarded as a necessary evil.
4. No policy on governmental matters. Local governmental office called when in doubt or in trouble.

Industrial Relations

PROGRESSIVE CONDITION
1. An executive vested with adequate authority formulates sound industrial relations policies and represents the company in labor negotiations.
2. An industrial relations program planned that minimizes labor turnover, builds employee morale and efficiency.
3. Program for effective selecting, testing, placing and training of all personnel.
4. Salary and wage rates equitable and fair for each job classification from common labor to top management; established by sound job evaluation methods.
5. Incentive plans for all levels of employees, based on an equitable measurement of performance.
6. Individual history and progress records for each employee; kept up-to-date for use as an inventory of qualifications.

AVERAGE CONDITION
1. Value of industrial relations realized, but authority and responsibility not clearly defined.
2. Industrial relations program not planned ahead, but ably administered; employee morale and efficiency average.
3. Employee selection not developed beyond separate formulas used by Office Manager for clerical help and by Employment Manager for all other help.
4. No job evaluation program. Wage rates increased under pressure. Top management and supervisory positions awarded largely on basis of seniority.
5. Incentive plans for some employees only.
6. Some records, but they are incomplete.

WEAK CONDITION
1. The industrial relations function is one of employment only, also coupled with other unrelated functions.
2. Little consideration given to industrial relations—employee turnover high.
3. No uniform procedure for applicant screening, placement and training. Original interviewing left to each department head.
4. Job rates fixed by personal opinion.
5. No incentive plan.
6. No records kept on individual employees beyond payroll requirements.

Finance

PROGRESSIVE CONDITION
1. Forecast of working capital and cash requirements for planned business volume and profis level.
2. Adequate reserves for replacement of obsolescent and depreciating assets— represented by earmarked liquid funds to the extent required.
3. Dividend policy consistent with sound, long-term financial program.

AVERAGE CONDITION
1. No forecast of working capital or cash requirements. Funds not always obtained or employed.
2. Depreciation reserves conditioned on allowable deductions for tax purposes only; not properly planned from a capital asset replacement point of view.
3. No definite financial or dividend policy.

WEAK CONDITION
1. Working capital and cash inadequate; credit policy lax. No forward planning.
2. Nominal reserves without due regard to actual value of assets; frequently used for purposes other than originally intended.
3. Financing dictated by immediate need for cash to meet pressing obligations.

Product Research

PROGRESSIVE CONDITION
1. Continuous research for improvement of products, equipment and methods of manufacture; for the development of new products and markets.
2. Efforts thoroughly planned and highly organized under expert supervision with qualified personnel.
3. Close cooperation with merchandising and sales, manufacturing, methods and tool engineering, to insure market acceptance, proper manufacturing facilities and a competitive production cost.

AVERAGE CONDITION
1. Research effort spasmodic; objectives not definite.
2. Research and development activities not well organized.
3. The company has a program for product research, design and engineering, but these activities are not carried out in cooperation with other divisions of the business.

WEAK CONDITION
1. No research. Product design and engineering inadequate.
2. No personnel qualified to conduct research and development.
3. Need for research and product engineering ignored.

Figure 7-1. How to Analyze Your Own Business—Continued

PROGRESSIVE CONDITION	AVERAGE CONDITION	WEAK CONDITION
II. PRODUCTION		
Procurement		
1. Purchasing of all materials through competitive bids, in accordance with specifications, in quantities requisitioned by Production Control. Effective expediting procedures.	1. Purchasing function generally well handled. Lacks complete coordination with Engineering and Production Control. Fair expediting procedures.	1. Purchasing not completely centralized. Poorly coordinated with Engineering, Production Control and other departments. Poor expediting procedures.
Production Control and Scheduling		
1. Production completely planned and scheduled in accordance with Sales requirements and manufacturing facilities.	1. Production planned as to principal items. Scheduling of material and labor needs by department heads.	1. No central production control and scheduling. Production often dictated by need to keep men busy—resulting in unbalanced, excess inventories.
Plant Engineering		
1. Plant location determined by studies of material and labor supply and market location. Plant facilities arranged in accordance with production methods and processes. Maintenance and replacement of plant, equipment and facilities well controlled.	1. Plant not located as a result of economic study. Machinery and equipment layout not well correlated to material flow. Maintenance and replacement of plant and equipment loosely controlled.	1. Plant location determined by available building space. Machinery location and plant layout arranged with little regard to economical material flow or handling. No facility replacement program and poor control of maintenance.
Tool Engineering		
1. Tools developed, designed and tested to yield the lowest feasible manufacturing cost for each product. Tools efficiently maintained and controlled.	1. Tools well constructed but not designed to produce lowest manufacturing cost. Fair tool maintenance and control.	1. Tool engineering not well correlated with manufacturing processing to produce low product costs. Poor tool maintenance inventory control.
Methods Engineering		
1. Head of methods and process engineering capable of developing, improving, standardizing and simplifying manufacturing processes to reduce costs in cooperation with Factory and Engineering departments.	1. Separate Methods and Standards department. Full coordination with Manufacturing, Tool and General Engineering departments not maintained.	1. Methods worked out by various department heads—improvements slow. Poor records and little control of manufacturing processing.
Manufacturing		
1. High-quality low-cost production for all products obtained by use of modern machinery, good plant layout and material flow, with high labor efficiency attained with incentive pay and able supervision.	1. Material flow needs improvement. Machinery up-to-date. Costs not low in field; loose incentive rates for labor. Improved supervision needed.	1. Manufacturing not well planned or supervised. Machinery old, material flow poor, product quality fair. No incentive pay rates—supervision indifferent.
Quality Control		
1. Quality control segregated as a separate function. Efficient inspection program tailored to each product and used as aid to sales and manufacturing.	1. Quality control function not centralized. Inspection performed as a manufacturing necessity only, except when quality complaints are made by customers.	1. No separate quality control function except when complaints force extra precautionary measures. Inspection carried on independently by each department foreman.

Figure 7-1. How to Analyze Your Own Business—Continued

PROGRESSIVE CONDITION	AVERAGE CONDITION	WEAK CONDITION

III. DISTRIBUTION

Sales - Merchandising

PROGRESSIVE CONDITION	AVERAGE CONDITION	WEAK CONDITION
1. Sound sales program—based on known customer needs, market research and analysis—supported by good advertising and sales program	1. Sales program based on past customer experience. Market potential not known. Advertising not selective.	1. Sales coverage incomplete. Knowledge of competition limited.
2. Sales budgets classified by products, customers, salesmen, territories.	2. Sales total estimated but not budgeted by products to customers and territories.	2. No sales budgets.
3. Sound pricing based on standard costs, giving effect to all cost factors.	3. Price structure rigid. Accurate product costs not used in setting sales prices. Competition partially governs pricing.	3. Selling prices based on competition or what market will bear. Costs information not generally used in setting prices.
4. Profit or loss determined by salesmen, territories, customers and products.	4. No attempt to analyze gross and net profit by salesmen, products, customers.	4. No sales analyses.
5. Selective selling effort directed toward maximum profit possibilities.	5. Selling effort not directed toward best profit possibilities.	5. No selective selling program.
6. Trained sales force intelligently directed and compensated.	6. Salesmen closely supervised but training program inadequate.	6. Sales force not well trained nor supervised. Compensation not comparable to competitors.
7. All sales records maintained currently.	7. Sales records not always maintained on a current basis.	7. No sales records beyond orders booked and sales billed.

Warehousing

PROGRESSIVE CONDITION	AVERAGE CONDITION	WEAK CONDITION
1. Warehousing program designed to meet customer delivery requirements—determined by a study of competitive market conditions.	1. Warehousing designed to service large established sales centers only. Not determined by analysis of competitive market conditions.	1. Warehousing maintained at manufacturing location—no consideration given to competitive sales and market conditions.

IV. CONTROL

Standard Costs

PROGRESSIVE CONDITION	AVERAGE CONDITION	WEAK CONDITION
1. Cost system designed to reflect all variances between standard and actual costs.	1. Cost accounting fairly accurate but not organized to provide standard cost information promptly.	1. No standard costs. Job costs inaccurate and uncontrolled.
2. Variances from standard performances supplied currently to management for corrective action (daily or weekly as needed).	2. Records and reports not best suited to control costs and expenses.	2. Cost information mostly estimated. Monthly profit and loss statements inaccurate.
3. Unnecessary accounting records eliminated—management control reports furnished as needed.	3. Many records, reports and statistics maintained that are not useful as a tool of management.	3. Some records and reports prepared; have no practical advantage.
4. All control records and costs integrated with standard costs.	4. Records unrelated to control; therefore, of little assistance.	4. Production records required for suitable cost control not maintained.
5. All estimates for product pricing based on standard costs; guesswork is eliminated; loss of volume or profit is indicated.	5. Estimates not checked against actual cost.	5. Estimates determined by past performance and competition.
6. The effect that sales mixture and product selling prices have on the total company profits-picture at varying operating levels is known at all times.	6. No knowledge of the effect on total business profits of individual product or order pricing.	6. Profit or loss estimated monthly; verified and adjusted annually to inventory; no profit or loss known by product breakdown.
7. Effect of additional volume on cost and profit is easily determined. Break-even points determined.	7. Effect of additional volume on cost and profit not easily determined. Break-even points not determined and their value underestimated.	7. Additional volume usually authorized to keep plant busy without knowledge of effect on cost and profit. No knowledge of sales mixtures or break-even point.

Figure 7–1. How to Analyze Your Own Business—Continued

PROGRESSIVE CONDITION	AVERAGE CONDITION	WEAK CONDITION
Budgetary Control		
1. Budgetary control of all expenditures based on flexible performance standards equitably established by operating levels.	1. Budget structure rigid; ratios of expense to sales based on past performance, not on predetermined, flexible performance standards.	1. No attempt made to budget or forecast performance.
2. Sales budget by products, salesmen, customers, territories, based on market analyses.	2. Sales budget by products, salesmen, customers and territories—based on past sales performance only.	2. No sales budget. No "quotas" for salesmen. No program.
3. Knowledge and control of the effect of selling price changes on budgeted amount of total net profits.	3. No centralized control of selling prices within limits of predetermined profit requirements.	3. No established pricing policy. Cost estimates ignored where considerable volume is involved. Effect of cutting prices to meet competition not projected in terms of lost profits.
4. Daily, weekly or monthly reports on the performance of all departments controlled through: (a) standard or budgeted performance and (b) variance from standard performance.	4. Divisional accounting reports periodically exhibited: (a) comparison of current with past periods; (b) no standards, therefore no comparison of actual results with what should have been accomplished, and no analysis of the causes of variations.	4. No budgets; no broad long-term planning. Policies vacillating because not founded on complete comparative information and thorough analysis.
Accounting		
1. Procedures, records, forms, reports designed with a view to producing required information at lowest cost.	1. Accounting fairly comprehensive, accurate, prompt and well managed—some written procedures.	1. Accounting accurate from bookkeeping standpoint, but generally "old-fashioned" and incomplete.
2. Accounting data supplied promptly, in a form best adapted to its use by management.	2. Accounting data not adequate in comparison with most modern conceptions of control by standards.	2. Accounting not highly regarded as a tool of management
3. Modern accounting equipment used effectively in preparation of necessary information and reports.	3. Accounting machines used but not adaptable to modern methods.	3. Accounting equipment antiquated, cumbersome and wasteful.

Source: Small Business Administration, "How to Analyze Your Own Business," Management Aid No. 46, reprinted September 1966.

and weaknesses lie. A chief executive may not be fully aware of how his company shapes up. He can, however, start at the same point at which an outside consultant would start: analyzing each aspect of his business. For this purpose, the Small Business Administration has prepared an excellent checklist (Figure 7–1).

Before using the checklist the chief executive should ask himself the following questions to check the effectiveness of his present organizational structure. If some of the answers are "no," he should then make improvements in the organization before attempting to make the improvements indicated by his answers to the checklist:

1. Is there a willingness on the part of top executives to delegate authority and responsibility, and to build up the skills, capacity and experience of others?

2. Are key men concerned mostly with policy or equally important activities, and not immersed in day-to-day routine?

3. Have functions and responsibilities of key men or departments been clearly defined?

4. Does every individual know to whom he reports and who reports to him, so that the lines of authority are not tangled and confused?

5. Are basic functions grouped logically, and is there an attempt to take advantage of specialization?

6. Is there a definite organizational plan, chart or write-up, and has it been studied recently?

Having answered these organizational questions and having initiated prompt action, the top manager is then in a position to analyze his business as to whether it is weak, average or progressive in each area. The answers will tell him where he needs to take action, and also what points to emphasize in presenting his company in its most favorable light to potential merger partners.

Every company has its problem areas, no matter how much it tries to improve itself. How should an owner answer questions regarding his weaknesses, which may come up during merger discussions? Obviously it would strain his credibility if he tried to deny areas that are, in fact, weaknesses. The best course of action is to treat the problem realistically, in terms of what he would do to correct the situation, the cost involved and the advantages expected to result from these improvements.

The capstone of this basic preparation for merger is a projection of financial statements. A form useful for this purpose is shown in Figure 7–2. With it and his backup materials, a chief executive should be ready to talk merger on favorable terms with any buyer.

Figure 7–2. Projection of Financial Statements

SUBMITTED BY

				ACTUAL	PROJECTIONS			
		SPREAD IN HUNDREDS ☐	DATE					
		SPREAD IN THOUSANDS ☐	PERIOD					
	1	NET SALES						
P	2	Less: Materials Used						
R	3	Direct Labor						
O	4	Other Manufacturing Expense						
F	5							
I	6	COST OF GOODS SOLD						
T	7	GROSS PROFIT						
	8	Less: Sales Expense						
and	9	General and Administrative Expense						
	10							
L	11	OPERATING PROFIT						
O	12	Less: Other Expense or *Income* (Net)						
S	13	Income Tax Provision						
S	14							
	15	NET PROFIT						
	16	CASH BALANCE (Opening)						
C	17	Plus RECEIPTS: Receivable Collections						
A	18							
S	19							
H	20	Bank Loan Proceeds						
	21	Total						
P	22	Less: DISBURSEMENTS: Trade Payables						
R	23	Direct Labor						
O	24	Other M'fg Expense						
J	25	Sales, Gen'l and Adm. Exp.						
E	26	Fixed Asset Additions						
C	27	Income Taxes						
T	28							
I	29	Dividends or Withdrawals						
O	30							
N	31	Bank Loan Repayment						
	32	Total						
	33	CASH BALANCE (Closing)						
	34	ASSETS: Cash						
	35	Marketable Securities						
	36	Receivables (Net)						
	37	Inventory (Net)						
	38							
	39	CURRENT ASSETS						
B	40	Fixed Assets (Net)						
A	41							
L	42							
A	43							
N	44	Deferred Charges						
C	45	TOTAL ASSETS						

Continued

Figure 7–2. Projection of Financial Statements—Continued

E	46	LIABILITIES: Notes Payable—Banks								
	47	Trade Payables								
S	48	Income Tax								
H	49									
E	50									
E	51	Accruals								
T	52	CURRENT LIABILITIES								
	53									
	54									
	55									
	56	CAPITAL STOCK ⎫ Net Worth for								
	57	SURPLUS ⎭ Partnership or Individual								
	58	TOTAL LIABILITIES AND NET WORTH								
	59	WORKING CAPITAL								

Source: Reprinted by permission of Robert Morris Associates and Cadwallader & Johnson, Inc., Chicago, Ill.

COMMENTS

SALES FORECAST

Consider (1) previous years business; (2) estimates of (a) Sales Department (b) Purchasing Department (c) Production Department; and (3) allowances for (a) economic outlook (b) government regulations (c) market (d) styles (e) peak periods. Space for comments at right.

INDICATE FACTORS USED IN PREPARING PROJECTION

    1. Average receivable collection period in days   _____
    2. Inventory Turnover in Days   _____
    3. Trade Payable Turnover in Days   _____
    4. % Federal Tax to Profits before Tax   %_____
    5. Depreciation per Year $_____
    6. Total Officers' of Partners' Compensation per month $_____

SUGGESTIONS FOR PREPARATION OF PROJECTION

Other Estimates needed for each period of the Projection are underlined below.
Blank lines in Projection are to accommodate unusual items of significance.
References to the Divisions of the Projection are abbreviated as follows:

                        Profit and Loss Statement   is PL
                        Cash Projection Receipts   is CR
                        Cash Projection Disbursements is CD
                        Balance Sheet Assets   is BA
                        Balance Sheet Liabilities   is BL

In the first column, record the actual PROFIT AND LOSS STATEMENT and BALANCE SHEET of date immediately prior to projection period.

In each subsequent column covering a projection period (month, quarter, etc.):

    1. Enter on date line, projection period covered and ending date thereof.

    2. Complete PL, recording NET SALES, less all discounts and allowances; showing costs and expenses as indicated. *Compute NET PROFIT OR LOSS.

Figure 7–2. Projection of Financial Statements—Continued

3. Record in CD on lines indicated, PL entries for DIRECT LABOR. OTHER MF'G EXPENSE, SALES, GENERAL and ADMINISTRATIVE EXPENSE and OTHER EXPENSE, less depreciation expense included therein. Record in CR, OTHER INCOME (PL).

4. Combine FIXED ASSETS (per prior column BA) and fixed asset additions, subtract depreciation expense and enter result in FIXED ASSETS (BA). Record cost of fixed asset additions in CD.

5. Combine INCOME TAX PROVISION (PL) with INCOME TAXES (per prior column BL), subtract payment of income tax and record result as INCOME TAXES (BL). Record income tax payment in CD.

6. Combine NET PROFIT or LOSS (PL) with SURPLUS or NET WORTH (per prior column BL), subtract DIVIDENDS OR WITHDRAWALS, record result as SURPLUS or NET WORTH (BL). Record DIVIDENDS or WITHDRAWALS in CD.

7. Record CASH (per prior column BA) as CASH BALANCE (opening) (CR).

8. Combine RECEIVABLES (per prior column BA) with NET SALES (PL), allocate resulting total between RECEIVABLE COLLECTIONS (CR) and RECEIVABLES (BA) per average collection period (Factor 1 above).

9. Combine TRADE PAYABLES (per prior column BL), with cost of material purchased (less discounts), allocate resulting total between TRADE PAYABLES (CD) and TRADE PAYABLES (BL) per turnover of payables (Factor 3 above).

10. Combine INVENTORY (per prior column BA), cost of materials purchased (less discounts) and DIRECT LABOR and OTHER M'FG EXPENSE (PL), subtract COST OF GOODS SOLD (PL), record result in INVENTORY (BA).

11. Review all items in prior column Balance Sheet (except CASH and NOTES PAYABLE—BANKS) for which no entries have been made in present period BALANCE SHEET. If there is no change in these items, transfer to present period BALANCE SHEET. If items are changed, reflect changes through CR or CD. Carry deferred charges (BA) and accruals (BL) without change.

12. Foot CASH PROJECTION: If cash deficiency indicated, enter amount to adjust in BANK LOAN PROCEEDS (CR); Combine this adjustment with NOTES PAYABLE—BANKS (per prior column BL) and enter as NOTES PAYABLE—BANKS (BL); if excessive cash is indicated, and NOTES PAYABLE—BANKS (per prior column BL) appears, provide BANK LOAN REPAYMENT (CD); reduce NOTES PAYABLE—BANKS (per prior column BL) by this provision, entering results as NOTES PAYABLE—BANKS (BL). Refoot CASH PROJECTION and enter resulting CASH BALANCE (closing) as CASH (BA).

13. Foot and balance BALANCE SHEET.

For Manufacturer projecting substantial increases or decreases in inventory during projection period. Enter as title on Line No. 5(PL).

"INCREASE OR DECREASE IN WORK IN PROCESS AND FINISHED INVENTORIES"—record increase in red, decrease in black.

Figure 7–3. Accounting Magic

All "In conformity with generally
Company B's Profits

	Company A Col. 1	Use of Fifo in Pricing Inventory Col. 2	Use of Straight-line Depreciation Col. 3
Sales in units	100,000 units		
Sales in dollars	$ 100 each		
	$10,000,000		
Costs and expenses:			
Cost of goods sold	$ 6,000,000		
Selling, general and administrative	1,500,000		
LIFO inventory reserve	400,000	$(400,000)	
Depreciation	400,000		$(100,000)
Research costs	100,000		
Pension costs	200,000		
Officers' compensation:			
Base salaries	200,000		
Bonuses	200,000		
Total costs and expenses	$ 9,000,000	$(400,000)	$(100,000)
Profit before income taxes	$ 1,000,000	$ 400,000	$ 100,000
Income taxes	520,000	208,000	52,000
	$ 480,000	$ 192,000	$ 48,000
Gain on sale of property (net of income tax)	—	—	—
Net profit reported	$ 480,000	$ 192,000	$ 48,000
Per share on 600,000 shares	$ .80	$ .32	$ .08
Market value at:			
10 times earnings	$ 8.00	$ 3.20	$ .80
12 times earnings	9.60	3.84	.96
15 times earnings	12.00	4.80	1.20

(   ) Denotes deduction.

Explanation of Columns 2 to 7, inclusive

Column	Company A	Company B
2.	Uses Lifo (last in, first out) for pricing inventory	Uses Fifo (first in, first out)
3.	Uses accelerated depreciation for book and tax purposes	Uses straight-line
4.	Charges research and development costs to expense currently	Capitilizes and amortizes over five-year period

(If R&D costs remain at same level, the difference disappears after five years. The difference of $80,000 in the chart is in the first year, where A expenses $100,000, and B capitalizes the $100,000 but amortizes 1/5.)

Figure 7–3. Accounting Magic—Continued

accepted accounting principles"
Are Higher Because of

Deferring Research Costs over Five Years Col. 4	Funding Only the Pensions Vested Col. 5	Use of Stock Options for Incentive Col. 6	Including Capital Gain in Income Col. 7	Company B Col. 8
				100,000 units
				$100 each
				$10,000,000
				$ 6,000,000
				1,500,000
				———
				300,000
$(80,000)				20,000
	$(150,000)			50,000
				200,000
		$(200,000)		
$(80,000)	$(150,000)	$(200,000)	———	$ 8,070,000
$ 80,000	$ 150,000	$ 200,000	———	$ 1,930,000
42,000	78,000	104,000	———	$ 1,004,000
$ 38,000	$ 72,000	$ 96,000	———	$ 926,000
———	———	———	$150,000	150,000
$ 38,000	$ 72,000	$ 96,000	$150,000	$ 1,076,000
$ .06	$ .12	$ .16	$ .25	$ 1.79
$.60	$1.20	$1.60	$2.50	$17.90
.72	1.44	1.92	3.00	21.48
.90	1.80	2.40	3.75	26.85

Column	Company A	Company B
5.	Funds the current pension costs—i.e., current service plus amortization of past service	Funds only the present value of pensions vested
	(Difference in pension charges might also arise where, as in the case of U.S. Steel in 1958, management decides that current contributions can be reduced or omitted because of excess funding in prior years and/or increased earnings of the fund or the rise in market value of the investments.)	
6.	Pays incentives bonus of officers in cash	Grants stock options instead of paying cash bonuses
7.	Credits gains (net of tax thereon) directly to earned surplus (or treats them as special credits below net income)	Includes such gains (net of income tax thereon) in income

MAGIC WITH NUMBERS

How one company can sell for more than twice another, although both have the same sales, cost of goods sold and selling expenses, is shown by Mace and Montgomery, based on a tabulation by Mr. Leonard Spacek, partner of Arthur Andersen and Company.[3] "Accounting Magic," Figure 7–3 shows how this is done.

If each company were sold on the basis of ten times earnings, and without any further accounting adjustments, *Company A's owners would receive $4.8 million, while Company B's owners would receive almost $11 million!*

In negotiations, an acquiring company would probably recognize the disparity in effects between the accounting practices used by the two companies. But it is easier for a selling company to withdraw from an overstated position than to advance from an understated one. The owner contemplating a merger might consider adopting "generally accepted accounting principles" that will recast his operating statements in a more favorable light.[4]

KNOWLEDGE OF A COMPANY'S VALUE

The vice-president for acquisitions approached two partners about selling their company. "What do you really want for your company?" he asked.

The senior partner replied, "Oh, I don't know—maybe half a million for me and half a million for my partner."

The partner's thinking, he divulged later, went something like this: They had started their business sixteen years earlier with $150 each. They had worked hard, lived modestly and had reinvested most of their earnings in the company. To go from the standard of living they were accustomed to, on about $20,000 a year, to what they could afford if they sold the company, it made little difference whether they each got a million or half a million for it. The proceeds would more than satisfy their needs for the rest of their lives. In other words, their criterion for price was, primarily, what they wanted to satisfy their needs in retirement.

Fortunately for them, the deal did not go through. The company was worth considerably more than their offhand estimate.

Value is in the eyes of the beholder; it is seen differently by different would-be acquirers, based on their various needs and assessments. It is also a matter of how the owner looks upon his company. In many cases a company that has a favorable earnings record is inclined to overvalue itself. In other cases, as the preceding, the owners may undervalue their firm. In any event, one way of avoiding a too-low

---

[3] Mace and Montgomery, *op. cit.,* 178–179: Reproduced through courtesy of the Harvard University Press and Mr. Spacek.

[4] Pricing inventory on the basis of FIFO (first in, first out) rather than LIFO (last in, first out) favors the selling company because FIFO values inventories at current replacement costs which, due to inflation, are higher than the "in" costs. LIFO tends to produce a lower inventory value relative to FIFO because "out" costs are practically the same as "in" costs.

price estimate is to consider the company's value from different points of view, based on the factors discussed in Chapter 3.

The owner-manager should also assess his value as a manager, apart from the value of the company itself. His continuing service may be a major asset of his company, both as to its selling price and as to its future value to an acquirer. On the other hand, if he has built up too much dependency on himself, an acquirer may consider the company less valuable because of its reliance on the health, intentions and motivation of a single person.

CONCLUSION

Putting its best foot forward—physically and in its operating results—is important in enhancing a company's desirability to an acquirer. Improving the basic value of a company takes one to three years, with constant attention to profit planning and both short-term and long-term results. The owner should also make a realistic appraisal of the company's value to avoid a too-low or too-high asking price; and he may want to consider the alternative effects of varying accounting practices.

But price is also a function of determining what potential buyers are most interested in the company, and of expertise in negotiations. The owner may find that professional help in effecting the best merger deal is well worth the investment.

CHAPTER **8**

THE MERGER BROKER

Most sellers of closely-held companies are very good at running their business operations but are woefully inept at the business of selling their business. Some of the problem areas the would-be seller encounters are:

- Where to get the *time* required to prepare the company for selling, keep pace with normal operations and go about the specialized business of a merger.

- How to maintain secrecy as to the company's availability, so that employees, competitors, bankers, suppliers and customers will not react adversely.

- How to determine who might be interested in acquiring the company.

- How to evaluate what price the business could bring.

- How to evaluate a proposed transaction in view of the tax, financial, operational and investment aspects involved.

- How to prepare for and conduct negotiations.

But the biggest problem, with acquisition-minded companies becoming increasingly wary of what is available in merger partners, is how to avoid the implication, "What's wrong with his company, if he's so interested in selling?" Even if nothing is "wrong" with the company, a potential acquirer will invariably be searching for an answer to "what is the *real* reason he wants to sell?"

Thus many owners of closely-held companies believe that, with the stakes so high, they need all the help they can get in selling their businesses. Even if an owner-manager has a good understanding of the merger transaction, however, the matter

97

of enhancing the value of his company will require that he give extra attention to what he does best: running his business.

The individual particularly qualified to handle and coordinate the business aspects of merging is the merger broker: part businessman, part merger expert and part swoose—a well-rounded individual who is a combination lawyer/engineer/accountant/marketeer/financial advisor.

WHAT A BROKER IS—AND IS NOT

Before looking at who a broker is and what he does, it is necessary to define the terms that describe him. Sometimes he is alternately called a mergant (merger merchant), a merger packager, merger intermediary, merger consultant or just plain merger broker. In dictionary terminology, these various titles describe an agent who buys or sells for a principal on a commission basis, without having title to the company.

From this definition it should be apparent that there is one thing the merger broker is *not:* a tipster or finder. Current literature not infrequently uses the two terms interchangeably, and there is no quicker way to raise the hackles of the merger broker. The better merger brokers are highly respected professionals whose integrity and record of service are their foundations for continuing success. After listening to a prominent broker describe his function, the vice-president of a major West Coast bank said, "Well, Pete, it is obvious that what you are really doing is a high-level consulting job on a contingency basis." This may be the most accurate description of all.

Who Are Brokers?

Who, then, are the merger brokers? The range of individuals and organizations is considerable: accounting firms, management consultants, commercial banks, investment banks and even the government, as well as firms whose fundamental business is serving as merger brokers.

The Federal government gets involved by having to find buyers, where necessary, to make divestiture orders stick. Both the Federal Trade Commission and the Justice Department have worked out brokerage arrangements necessary to enforce divestitures.[1]

Some commercial banks, seeing a potential loss of industrial customers via the merger route, are active mergants in order to be in a favored position in post-merger financing, and to match the merger services provided by competitors. Bank of America, Security Pacific and Chase Manhattan are among the banks who serve as merger brokers.

Investment bankers have also begun to play a major role as intermediaries, particularly for the larger companies who are party to a merger, for they see this

[1] "Finding Homes for Merger Orphans," *Business Week,* Dec. 9, 1967, p. 154–160.

service as a door-opener to subsequent underwriting business. The immediate fee potential is also worthwhile: Lehman Brothers received over $900,000 for five years' study and negotiations that culminated in the merger of American Home Products and Ekco Products Company. Other investment bankers and stock brokerage firms are also active in effecting mergers. Some are beginning to provide these services to smaller companies as well as to large ones.

CPA firms and management consultant organizations have become participants as merger consultants and brokers because of their inner knowledge of clients' goals, strengths and weaknesses, and because of their reputation for objective advice.

The head of a smaller company, looking for a reliable, disinterested merger broker, might check out each of these possibilities (banks, security brokers and professional merger organizations) for the expertise needed in the merger process. In most cases, however, he will find that a professional merger broker, whose business is not related to financing, loans, accounting or general consulting, is the one who will pursue his merger goals most aggressively and effectively. It is important for the seller to know with whom he is dealing, however. He should satisfy himself that the merger broker he selects is experienced, ethical and highly qualified before entrusting the merger transaction to him. The advantages of retaining such a broker, compared to possible loss through going it alone or with less than expert advice, are well worth the fees involved.

What a Broker Is NOT

A broker is definitely *not* a finder, although finding a buyer may be one of his functions; and a finder is definitely *not* a broker. It is important that the seller understand the differences between the two, and what he can expect from each.

Consider first the working arrangements. A broker serves as an agent for his principal, and has a fiduciary responsibility to the principal. He invariably has a contract covering the company's arrangement with him. This serves as a letter of authorization in his quest for the best buyer, on his client's behalf. The contract should include who is going to pay his commission, when the commission is payable (normally on closing of the merger agreement), the amount of the commission, whether the arrangement is exclusive or nonexclusive, and the term of the agreement.

A finder, on the other hand, may or may not have such an authorization (the better ones do), and he has no particular responsibility to the company he is representing. Unfortunately, there are many who misrepresent themselves as having authority to act as finders; this is easily determined, however, by asking to see this authorization. This latter type may take credit months or years after having casually dropped a tip, and although the tip was not in fact acted upon, a subsequent merger can give rise to their taking court action to receive payment for their "services." Often they are successful in achieving their real goal: an out-of-court settlement. A Forbes article hangs a name on these unethical individuals, and quotes an investment banking partner: "Every industry has its streetwalkers, and finders are the streetwalkers of this business!"[2] That finding can be remunerative is illustrated by two finders in Glen Alden Corporation's acquisition of Schenley Industries, who claimed a $5 million fee in 1968.

The main difference between a broker and a finder, aside from ethics, authorization and skill required, is in the services each provides. A finder or tipster's function is limited to finding a merger partner—period. It is worth repeating that there is a high percentage of finders who are completely ethical and sincere. But there are others who may tell a potential buyer that they have found the seller he needs, and then proceed to find a seller and tell him they have found the buyer for him (if they subsequently merge, he may claim a fee from each). Two companies in Chicago, for instance, competitors with no intention of merging with each other, discovered they were being set up as potential merger partners by a finder.

The functions of the merger broker are much more inclusive than those of the finder and require years of business experience and skill in providing the services needed by his client.

THE MERGER BROKER: WHAT HE DOES

One merger broker says, "Basically, I don't do anything that the ordinary, intelligent child of three can't do, with thirty years' experience." The competent broker has had years of experience in business, understands the problems of both buyer and seller, and draws heavily on this experience in effecting a workable, satisfactory merger.

Primarily he performs four functions: (1) analysis of the seller's business; (2) search for likely buyers; (3) negotiations on his client's behalf; and (4) assistance in integrating the two firms. As a result of his skill, and considerable time, one mergant was able to perform these functions to the extent that the buyer and seller met for only ten minutes prior to closing.

The typical sequence of events starts with a call to the broker, from a consultant or perhaps a CPA, saying that the president of a closely-held company is thinking about selling. The broker calls on the president of the company and they discuss his reasons for selling. During the discussion the broker, knowing the importance of the personal factor in selling a closely-held firm, asks what the seller's five-year objectives are. This gives the broker and the owner a "handle" on how to accomplish these goals; if the individual wants to retire, for instance, he may want to diversify his investment or get a more liquid estate, and this may point toward a merger for securities or equity in a more widely-held company. Or, if the owner wants cash for his purposes, this indicates that the buyer should be in a relatively strong liquid-assets position.

---

[2] "The Streetwalkers," *Forbes,* Feb. 15, 1967, p. 42.

Through experience, the same broker asks whether the owner has discussed his intentions of selling with his wife, because he has found that when the couple face up to seeing each other for most of the day, every day, occasionally they would both prefer that he continue working!

Another aspect that is discussed is what the owner expects the sale of his company to bring. If the owner's price is too high compared to that of other companies in the same industry, the well-qualified broker will often advise the owner, tactfully but directly, "Frankly, on the basis of earnings, assets, potential and the going price/earnings multiple in your industry, I can't get it for you." Through experience he has found that it saves his time, and the seller's hopes, if he is frank at the beginning as to the price the company would bring.

He might also advise the owner to be prepared for two shocks, even though neither may occur. One is when the owner finds out what his company is really worth; and the other is the day he realizes the company is no longer his. If the owner has faced these facts, then the two are ready to talk in depth about the aspects of a merger or sale. The broker can also prime the seller on future relations with his new employer if the seller wants to remain with his organization.

The Brokerage Contract

After this discussion the broker presents the standard exclusive agreement (Figure 8-1) for serving as the firm's merger broker, and suggests that the company's lawyer review it. The president should also confirm that he will give the broker full access to company information, and that any changes, good or bad, will be immediately made known to the broker.

For the protection of both the seller and the broker, a written contract should always be used to record their agreement. This may be an exchange of letters, in which the broker outlines the agreement and the seller concurs to the arrangements stated; or it may be a contract as such. Most brokers and sellers prefer a formal contract, and this is highly recommended to avoid possible problems resulting from an implied contract. Figure 8-1 shows the terms usually included. In negotiating a selling price, the seller should recognize, of course, that out of the payment he receives he in turn is obligated to pay the broker.

At the conclusion of the brokerage arrangement, either through accomplishing a merger in accordance with the contract or through expiration of the terms of the agreement, the seller should send a letter discharging the broker from further efforts on his behalf. This is primarily for the seller's protection so that the transaction will be fully documented; but it also releases the broker, in case there may be some questions as to the availability of his services, as in post-merger integration.

Regarding this agreement, the reasons for an exclusive are similar to those of a capable real estate agency. First, the broker can afford to devote more intensive time

Figure 8–1

FRANK, KIMBALL, PARSONS & DAUM, INC.

LOS ANGELES · HOLLYWOOD · SAN FRANCISCO

October 25, 1965

PERSONAL AND CONFIDENTIAL

Dayton Manufacturing Company
624 East Vail Avenue
Montebello, California

Attention: Mr. George Davis, President

Gentlemen:

This letter confirms our understanding and states the terms of
our agreement concerning the services we will perform in connection with your desired sale or exchange of all or a portion of
the capital stock or assets of DAYTON MANUFACTURING COMPANY.

It is understood that the purchaser must be completely acceptable to you and that the terms and conditions are completely
subject to your approval. Our company shall act as your sole
and exclusive agent in connection with the aforesaid transaction for a period from the date of your acceptance of this
letter. This agreement, however, may be terminated at any
time by either party upon ninety days written prior notice
to the other party.

We will prepare an analysis of your company and will provide
a comprehensive presentation tailored for each qualified buyer
whom we have selected after conducting a discreet search. You
agree to make all of your company records available to us and
further agree to disclose immediately to us any material change
in the company at any time prior to closing. We agree to hold
any data or information of a confidential nature received from
you in the strictest confidence and to use it solely in our
efforts on your behalf. Negotiations with prospective purchasers
will be handled by our organization, and inquiries which come
directly to you should be referred to us.

In consideration for our services, you shall pay us a commission
in accordance with the attached Commission Schedule which is a
part of this agreement. Our commission shall be deemed earned
if an agreement is reached with any purchaser introduced by us
provided, however, that the transaction shall be agreed upon and
signed during the period of this agreement or within a period
of one year from the date of termination of this agreement and
any extensions thereof. Our commission is due and payable no
later than the closing of any such transaction.

Figure 8-1 Continued

FRANK, KIMBALL, PARSONS & DAUM, INC.

Dayton Manufacturing Company   October 25, 1965
Page 2

If the preceding is your understanding of our agreement and
meets with your approval, please signify so by dating and
signing the attached copy of this letter and returning it to
us, whereupon this letter shall constitute a binding agreement between us.

                         Very truly yours,

                         FRANK, KIMBALL, PARSONS & DAUM, INC.

                    By _____

ACCEPTED AND AGREED THIS 2̲6̲ day of _O̲c̲t̲o̲b̲e̲r̲_, 19̲6̲5̲.

                         DAYTON MANUFACTURING COMPANY

                    By _____
                    Date _1̲0̲-̲2̲6̲-̲6̲5̲_____

on behalf of the seller. Second, most buyers prefer to deal with a broker who has an exclusive. Finally, an exclusive protects the seller from the unscrupulous tipster or finder.

Analysis

After these arrangements are made, the broker performs an analysis on the company, defining the owners' goals, making a personal evaluation and drawing up a corporate profile based on operating statements and balance sheets for the past five years. The profile includes as much information as possible, without leaving a "fingerprint" as to the company's identity. This profile lists assets, five-year earnings and other financial and operating information on a *pro forma* basis; that is, it is recast by the broker (and so indicated in a footnote) to reflect earnings on a basis comparable to an "institutionalized" (widely-held) corporation. Earnings thus are usually higher than those reflected in the company's regular statements. In the case of a company that showed $9,000 profit in its most recent fiscal year, recasting in the form of a buying company reflected an after-tax profit of $50,000. The price later paid for the company was based on this higher figure. Figure 8–2 shows a company's operating statements for five years, and the same statements recast to conform to practices in large corporations.

The profile also includes a three-year projection of earnings, and a statement of the source and application of funds. A sample profile is shown in Figure 8–3.

During this analysis the broker will explain to the seller how an acquirer evaluates a merger prospect as well as other aspects of an acquisition. He also advises the seller that any buyer will come in with his own auditors for a thorough check, when the merger procedure reaches this point. Then, prior to conducting the search, he gets together with the president and frequently the entire board of directors and explains what he believes the company can be sold for, and why and how.

The Search

Having done this homework, the merger packager determines why the company is attractive and to whom, and conducts a search for a buyer. His criteria for the success of the merger, and for his own continuing success as a merger broker, is that both companies must gain from the transaction.

Knowing what acquisition-minded companies are looking for is one way of determining potential merger partners for his client (brokers who have been in the business for some time may use an automated system, including 700 or more acquisition-minded companies). Sometimes, though, a more effective way is using his own judgment and imagination as to who would be good for whom. Obviously this requires considerable knowledge and experience.

His search is helped by the fact that many acquiring companies depend on brokers in addition to, or instead of, their own internal search analysts. Figure 8–4 is an example. One active acquirer, Associated Products, Inc. of New York, actively works with more than 180 brokers, of whom 30 regularly bring in merger prospects.[3] With so many acquisition-minded companies actively searching for merger partners, this indicates the scope of the potential market that the broker must know on a continuing, daily basis.

Any individual acquirer's needs vary from time to time, however, depending on availability of their investigating personnel, financial ability and an occasional need to "burp" while digesting previous acquisitions. For these reasons, finding the right buyer may take as little as a week or as long as two years.

When his screening "kicks up" half a dozen or so likely merger partners for his client, the broker calls the president of a prospective acquirer and asks whether he is interested in receiving the selling company's profile. If so, and if the company is interested after reviewing the profile, the broker asks the seller for permission to reveal his company's identity. The broker then sends the actual (audited) performance statements and balance sheets in order that the prospective buyer can make his own pro forma analysis of the selling company.

Negotiations

The next step is for the broker to arrange a get-together between the buyer and seller, but on one condition: that there will be no discussion of price or terms. The reason for this condition is to give full rein to exploring the "fit" and "chemistry" of the two companies and their principals. (An incidental advantage of having a third party act as go-between is that he can often avoid the personality conflicts that might otherwise arise between the two chief executives.)

After this meeting the broker advises the buyer of the asking price. "If the price is clearly out of line," one broker says, "an experienced buyer won't even look at the company. But if the buyer shows an interest in spite of the price, I tell him, '*You* analyze it, and determine whether it's a company that fits into your picture. Then let's discuss it further.' "

Assuming the buyer comes back with an offer, the broker transmits it and a proposed agreement, with the terms proposed, to the seller for his consideration.

From this point the broker's role is one of bringing the two parties together: reducing the difference in price and terms, and suggesting alternate ways of accomplishing each party's objectives. Here again, experience and imagination are required.

Another requirement is time. Great amounts of time are required in meetings

[3] Allan Lynn, Vice President, Corporate Development, Associated Products, Inc., "How to Make Contact," in G. Scott Hutchison (editor), *The Business of Acquisitions and Mergers* (New York: Presidents Publishing House, 1968), p. 138.

Figure 8–2

FRANK, KIMBALL, PARSONS & DAUM, INC.

LOS ANGELES · HOLLYWOOD · SAN FRANCISCO

File 9028

Summary of Income Statements
(000 Omitted)

	1965	1966	1967	1968	1969
Income from Sales	$ 772	$ 830	$ 870	$1,397	$1,267
Cost of Sales	506	539	543	860	820
Gross Profit	$ 266	$ 291	$ 327	$ 537	$ 447
Operating Expenses:					
Selling Expenses	$ 124	$ 137	$ 147	$ 188	$ 163
General Expenses	58	66	71	76	89
Total Operating Expenses	$ 182	$ 203	$ 218	$ 264	$ 252
Net Operating Profit	$ 84	$ 88	$ 109	$ 273	$ 195
Other Income:					
Interest Earned & Miscellaneous	$ 5	$ 6	$ 5	$ 6	$ 7
Discounts Earned	1	2	2	3	3
Rental Income	4	4	4	4	4
Total Other Income	$ 10	$ 12	$ 11	$ 13	$ 13
Net Income Before Exec. Salaries	$ 94	$ 100	$ 120	$ 286	$ 208
Less: Executive Salaries	68	68	78	92	91
Net Income Before Taxes	$ 26	$ 32	$ 42	$ 194	$ 117
Less: Provision for Fed. Inc. Tax	8	11	14	86	50
Net Income for period	$ 18	$ 21	$ 28	$ 108	$ 67

This summary prepared from individual annual statements prepared by company's auditors.

600 EAST 8TH STREET · LOS ANGELES, CALIFORNIA 90014 ; 213 / 680-9850

Figure 8-2 Continued

FRANK, KIMBALL, PARSONS & DAUM, INC.

LOS ANGELES · HOLLYWOOD · SAN FRANCISCO

File 9028

Summary of Pro Forma Income Statements
(000 Omitted)

Gross Income	$ 782	$ 842	$ 881	$1,410	$1,280
Cost of Sales	506	539	543	860	820
Gross Profit	$ 276	$ 303	$ 338	$ 550	$ 460
Expenses:					
Selling Expense	$ 124	$ 137	$ 147	$ 188	$ 163
General Expense					
(Note 1)	91	101	111	116	129
Total Expenses	$ 215	$ 238	$ 258	$ 304	$ 292
Net Income Before Tax	$ 61	$ 65	$ 80	$ 246	$ 168
Less: Provision for Fed. Inc. Tax (Note 2)	$ 24	$ 26	$ 33	$ 120	$ 80
Income for Period	$ 37	$ 39	$ 47	$ 126	$ 88

Pro Forma Balance Sheet
December 31, 1969

Cash on Hand	$ 297	
Notes & Accounts Receivable-Net	212	
Inventories (Note 3)	321	
Real Estate (Note 4)	270	
Machinery & Equipment (Note 5)	92	
Other (Note 6)	101	
Total Assets		$1,293
Liabilities		
Accounts Payable	$ 3	
Dividends	15	
Accrued Expenses	38	
Fed. Inc. Tax Estimated (Note 1)	50	
Total Liabilities		$ 106
Net Worth		
Paid-In Capital	$ 189	
Surplus	998	
Total Net Worth		$1,187
Total Liabilities & Net Worth		$1,293

Note 1: Provides $40,000/year for general manager.
Note 2: Does not include 10% surtax.
Note 3: Valued at lower of cost or market.
Note 4: Valued at current market.
Note 5: Net sound value.
Note 6: Includes $55,000 cash value executive insurance policy.

Figure 8-2—Continued

FRANK, KIMBALL, PARSONS & DAUM, INC.

LOS ANGELES · HOLLYWOOD · SAN FRANCISCO

File 9028

Summary of Balance Sheets

(000 Omitted)

	1965			1966	
Current Assets					
Cash on Hand		$ 191			$ 172	
Notes & Accounts Receivable	$ 87			$ 94		
Less: Allow. Doubtful Accts.	4	83		4	90	
Interest Receivable		--			--	
Merchandise Inventory:						
Merchandise on Hand	$ 66			$ 88		
Labor in Process	35			36		
Finished Product	1	102		3	127	
Total Current Assets			$ 376			$ 389
Fixed Assets						
Land		$ 53			$ 53	
Building & Improvements		52			52	
Machinery & Equipment		59			72	
Automobiles		12			12	
Office Furniture & Fixtures		16			16	
		$ 192			$ 205	
Less: Allow, Depreciation		96			108	
Total Fixed Assets			96			$ 97
Other Assets						
Cash Value, Life Insurance (Net)		$ 46			$ 46	
Small Tools, Patterns		1			3	
Prepaid Expenses		6			5	
Patents						
Organization Expenses		1	54		1	55
Total Assets			$ 526			$ 541
Current Liabilities						
Accounts Payable		$ 3			$ 2	
Dividends Payable		10			10	
Accrued Expenses		18			20	
Fed. Income Tax (Est.)		8			11	
Total Current Liabilities			$ 39			$ 43
Net Worth						
Common stock, 1,000 sh., no par		$ 189			$ 189	
Retained Earnings	$ 290			$ 298		
Add: Net Income	18			21		
Less: Dividends	(10)	298		(10)	309	
Total Net Worth			487			$ 498
Total Liabilities & Net Worth			$ 526			$ 541

600 EAST 8TH STREET · LOS ANGELES, CALIFORNIA 90014 · 213 / 680-9850

Figure 8-2—Continued

```
                1967                          1968                          1969
              $    159                     $    187                     $    147
$    98                        $   166                      $   216
      4            94                 4        162                4          212
                                               4
$   110                        $   155                      $   143
     44                             37                            55
     11           165                2        194                23         221
                       $    418                    $    547                      $    580


              $     53                      $     53                     $     53
                    52                            67                           67
                    76                            91                           95
                    12                            13                            6
                    16                            16                           17
              $    209                      $    240                     $    238
                   117                           127                          131
                       $     92                    $    113                      $    107


              $     49                      $     52                     $     55
                     4                             5                            3
                     8                             8                           17
                     2                             2                            2
                     1          64                 1          68               1          78
                       $    574                    $    728                      $    765


              $      6                      $      5                     $      3
                    12                            15                           15
                    28                            15                           38
                    14                            86                           50
                       $     60                    $    121                      $    106


              $    189                      $    189                     $    189
$   309                        $   325                      $   418
     28                            108                           67
    (12)          325              (15)        418              (15)        470
                       $    514                    $    607                      $    659
                       $    574                    $    728                      $    765
```

Figure 8–3

FRANK, KIMBALL, PARSONS & DAUM, INC.

LOS ANGELES · HOLLYWOOD · SAN FRANCISCO

File 9028

Projected Statement of Profit
(000 Omitted)

	1970	1971	1972
Gross Income	$1,500	$1,750	$2,100
Cost of Sales	910	1,060	1,280
Gross Profit	$ 590	$ 690	$ 820
Expenses:			
Selling Expense	$ 210	$ 234	$ 280
General Expense (Note 1)	123	143	172
Total Expenses	$ 333	$ 377	$ 452
Net Income Before Tax	257	313	368
Less: Provision for Fed. Inc. Tax (Note 2)	$ 133	$ 155	$ 175
Income for Period	$ 124	$ 158	$ 193

600 EAST 8TH STREET · LOS ANGELES, CALIFORNIA 90014 · 213/680-9850

Figure 8–4

TRANSITRON
ELECTRONIC
CORPORATION

Subsidiaries:

PHALO CORPORATION
SHREWSBURY
MASSACHUSETTS

ELECTRA/MIDLAND
COMPANY
KANSAS CITY
KANSAS

THE CALTYPE
CORPORATION
LOS ANGELES
CALIFORNIA

FLORIDA
TRANSFORMER
COMPANY
DE LEON SPRINGS
FLORIDA

BETHCRAFT
CORPORATION
BETHLEHEM
PENNSYLVANIA

CONAIR, INC.
GLENDALE
CALIFORNIA

FORREST WAGNIERE
ENGINEERING
COMPANY
LOS ANGELES
CALIFORNIA

METCRAFT
CORPORATION
BALTIMORE
MARYLAND

TRANS/CIRCUITS INC.
FALLS CHURCH
VIRGINIA

LEMCO
LONDON
ENGLAND

TRANSITRON
ELECTRONIC
LIMITED
BERKSHIRE
ENGLAND

TRANSITRON
ELECTRONIC S/A
AMSTERDAM
THE NETHERLANDS

TRANSITRON
MEXICANA S/A
NUEVO LAREDO
MEXICO

TRANSITRON
ELECTRONIC
CORPORATION

Plants:

BOSTON/MELROSE
ANDOVER/WORCESTER
MASSACHUSETTS
LAREDO/TEXAS
ANTRIM/NORTHERN IRELAND
VERNON, FRANCE

CORPORATE
HEADQUARTERS
WAKEFIELD
MASSACHUSETTS

March 7, 1969

Kimball & Company
235 Montgomery Street
San Francisco, California   94104

Gentlemen:

Your professional associations place you in a unique position to
know of special situations in the acquisition field. Perhaps you
know of a closely held or privately held company which is available
or could be available for merger or acquisition.

The reason could be one of many: a sole owner wishing to convert his
holdings under tax-free circumstances, a family owned company in
which the younger generation does not want to take over, a profitable
small business which wants to expand to take advantage of market
opportunities but needs the resources of a large corporation to
accomplish its objectives, a company producing for a few local
customers which could be a winner with the addition of highly professional nationwide marketing, etc.

Your interest in guiding us to such a situation might be to help save
a friend's company, to protect owners' interests, etc. In any case
we would work directly with you to initiate negotiation.

Our interest in acquiring is to continue the rapid growth rate at
Transitron, a growth that has seen the acquisition of eight companies
in the past year and has put Transitron over the $100 million mark
in annual sales. Our subsidiaries retain their identities and their
managements, yet benefit through the financial flexibility, management
depth and broad facilities of a large corporation.

If you are aware of a special situation in which our interests and
yours can be put together for our mutual benefit, you are invited to
write to me in confidence to set up a personal meeting where we may
explore possibilities.

Sincerely,

Jerome J. Kutzen
Director of Corporate Development

with attorneys, CPA's and other representatives of each of the merger parties; in explaining the offer to the selling company's board and chief executives; in handling counterproposals; and in making sure each party understands the other's goals, offer, terms, and so on. Often the mergant is a high-level expediter, particularly in situations where both companies are well run and under no pressure to complete the transaction.

In negotiations (described more fully in Chapter 11) the experienced merger intermediary is particularly valuable, both because his experience gives the seller an expertise comparable to the buyer's, and because the broker can arrange for capable legal, accounting and other professional specialists. Having worked with a number of these specialists, he is in a unique position to draw on those who will be of most benefit to his client.

For the owners of the selling company, the broker will have made a good merger if they are happy with the price received and with the company to whom they have sold their business. The final merger agreement is a good one for both parties if each is satisfied but cannot be described as "contented" with the deal.

Post-Merger Integration

During the merger process the merger broker has become familiar with the operations and goals of both companies. Because of this he is in an excellent position to assist in integrating their operations. If he is asked to do so, he can be particularly helpful in allaying the qualms of personnel in the selling company, and in getting them off to a good start with their counterparts in the acquiring company.

FEE

And what does the merger broker receive for these services?

Some firms charge a flat 3 per cent on the sale, but the most standard basis is "5-4-3-2-1," where the broker receives 5 per cent of the first million dollars, 4 per cent of the next million and so on down to 1 per cent on the fifth million. From a firm that sells for $5 million, the broker would receive $150,000. On a sale of more than $5 million, the fee is usually negotiated on an individual basis.

This is not a small amount. However, considering the tremendous amounts of time involved in a merger transaction, the skill required in avoiding pitfalls and the fact that the broker can maintain secrecy and make contacts that the seller himself cannot, most sellers who have in the past used a good broker would do so again. And acquirers also recognize the value of the merger broker. Mr. Edward Hoffman, Vice President, Acquisitions, of Dyson & Kissner Corporation, says, "The best money we spend in a merger is the fee paid to a competent broker."

SUMMARY

The services provided by an experienced broker are:

- Ability to maintain the confidential nature of the proposed merger

- Knowledge of likely acquirers

- Taking the owner "off the hook" in selling his own company

- Objective appraisal of the company's strengths, weaknesses, likely selling price

- Skill and experience in negotiations

But perhaps the broker's main contribution is in saving the time of the selling company's chief executive, so that he can continue doing the job he does best: running a profitable business. If the seller uses the same care he would in hiring the services of any other professional, his merger broker will prove to be extremely valuable.

CHAPTER 9

LEGAL, ACCOUNTING AND OTHER SPECIALISTS

In mergers, as in girl-watching, this is the age of the specialist. Specialization in merger law, for instance, has advanced to this level: "By now [merger and acquisitions law] has become so complex that it is probably little understood by lawyers outside the country's four or five largest cities."[1]

A merger is several transactions from the seller's point of view. It is a business transaction requiring a specialist on mergers from the business standpoint; it is a legal transaction requiring a lawyer; it is a financial transaction requiring expert tax and accounting counsel; and it is an investment in cases where the seller receives equities or securities in the buyer's company, requiring skilled investment advice. Each of these fields has changed greatly in the last several years, and this fact underlines the key point in selecting merger specialists: There is no substitute for merger experience. To put it another way, the seller who relies on his regular law firm, CPA or other normal-business advisor has a good chance of going wrong. These regular advisors are concerned, after all, with application of their specialties to *on-going* business operations rather than to the sale or merger of a client's company. Also, because they would be losing a client via the merger route, they actually have incentive *not* to see a merger transaction go through. Lawyers, particularly, can effectively kill a deal.

SELECTING THE SPECIALIST

Selecting specialists with experience is almost mandatory when one considers the expertise available to the acquirer: his own highly skilled legal and accounting personnel, business brokers, management consultants, commercial bankers, investment bankers,

---

[1] Joseph Poindexter, "The Cool, Creative Company Lawyers," *Dun's Review*, March 1969, p. 35.

appraisers and insurance counselors and others, *each of whom spends all or most of his time looking after an acquirer's interests.*

The seller does not need to match the buyer's team man for man, of course, in either numbers or in specific skills, although both parties need business, legal and accounting counsel. The fundamental purpose of additional specialists is to probe the value of what their principal—buyer or seller—will receive in the transaction.

One individual often overlooked by the seller, because most sellers fail to recognize they are making a major investment, is the investment counselor. Except in transactions that are solely for cash, this individual is the key advisor on the quality and merits of securities that may be offered by the acquirer.

How does the seller go about selecting his broker, attorney, CPA and investment advisor? And how can he select experienced merger specialists without offending his normal-business attorney, banker and CPA? One way is to ask his regular advisor, for instance his lawyer, to recommend an experienced merger lawyer. Often, however, this regular advisor may not know merger specialists. In this case the owner is advised to select an experienced merger broker and to ask his recommendations. A broker, having worked with merger specialists, will usually suggest three or four qualified merger attorneys and a similar number of CPA's so that the client may make his own choice from among these well-qualified professionals. If this kind of experience is not available to the seller, either he or the merger broker should make sure the legal, accounting and other specialists clearly understand what is expected of them.

As in any business arrangement, of course, the owner should ascertain ahead of time what the specialist's fees will be and the extent of the services to be provided by him. (Most experienced merger lawyers charge by the hour, the same as corporate-general attorneys do.) On mergers, the experienced specialist can work much more quickly, and more thoroughly, than his normal-business counterpart; hence the merger specialist's total fee is likely to be lower. In one case a seller paid his merger attorney $5,000 for two weeks' work. But another seller retained his regular attorney in a similar merger and paid more than $10,000 in legal fees for a less professional effort spread over so many weeks that the attorney almost killed the deal.

More important than the matter of fees, though, is the added value received in negotiations through the seller's having chosen professionals in mergers. And, by obtaining the best of professional advice, the seller will probably avoid the post-merger blues when he wonders, too late, whether he might have gotten a better deal if he had had better counsel.

LEGAL COUNSEL

The vice-president and legal counsel for a diversified corporation says, "If there is an important development in law in the last twenty years, it is acquisitions and mergers.

The aggregate of all the classic law in law school texts is overshadowed by the last five years' development in acquisitions law."[2]

Basically, the job of the attorney is to translate the principals' business arrangement into proper legal form, and to provide the necessary protection and warranties for his client. The lawyer may find that the business arrangement cannot be concluded exactly as the two principals have outlined; but if he is creative and ingenious, he can find legal ways and procedures to achieve the same results.

The lawyer's attitude has a great deal to do with how much he contributes to the merger transaction. First, he must recognize, as experienced merger lawyers do, that his function is not to try to revise the businessmen's arrangement, but to reduce it to contractual form. Second, it takes a creative approach—"here's how we can do it"—rather than a "here's why you can't do it" attitude. The lawyer who has little experience in mergers is likely to be so cautious in stepping into this unfamiliar field that he may try to change the basic agreement, or make it so safe for the seller that the transaction just will not close. The company's regular attorney, again, is probably *not* the man for this job.

A lawyer, particularly, can block a merger if he is opposed to it. Even when he is not opposed to it, he might effectively cause the scrapping of what could have been a mutually beneficial merger. In one instance a seller's lawyer mistakenly advised the buyer's attorney that an intermediary had no contract to represent the seller when, in fact, the intermediary did have such a contract. The buyer, who had made a deposit on the purchase, consequently asked for a return of his deposit. Another purchaser was later found and the deal was completed, but the seller's profit picture had meanwhile deteriorated, and this was reflected in the lower price paid.

Lawyers themselves recognize that some of their colleagues become overly intent on protecting all matters, great and small, of a client's interests. For this reason experienced lawyers and acquirers alike concur that lawyers should not be brought into discussions too soon. It is important first that the two businessmen have a chance to explore the business benefits to be gained through a merger.

The more the seller's lawyer knows about the business, its tax matters and contingent and other liabilities, the better he will be able to effect a satisfactory agreement for the seller. Due to previous experience, most acquirers now present an extensive list of warranties for the seller to comply with; these pertain both to business practices and to the condition of the company at the time it is sold. The lawyer for the seller should insist that there are no unnecessary warranties binding his client, and knowledge of the company's status is necessary in order to do this.

The stack of legal documents at the time of closure may be a foot high: bylaws, minute book, corporate seal, authorization to conduct business in the state, the merger contract itself, opinions of patent counsel and corporate counsel, real estate ap-

---

[2] Poindexter, *op. cit.*, quoting James Kilbridge of Walter Kidde & Co.

praisals, warranties of both parties, general provisions, resignations of officers and directors of the selling company, certificates regarding ownership, assignment of patents, and so forth. It is apparent that the owner of the selling company could incur a very heavy expense paying for the merger education of a lawyer who does not regularly practice merger law.

THE CPA

Just as the legal counsel should investigate the company's legal framework and various actual and implied contracts in order to talk knowledgeably with the buyer's attorneys, so should the company's CPA make a thorough audit. This has the purpose of preparing the company's position for negotiations.

The audit should be addressed to three fundamental questions:

- What is the profit and loss picture, and is it fairly represented by the company's records?

- What are the true value and nature of the firm's assets and liabilities, and are they correctly and completely shown by historical and pro forma balance sheets?

- How do the accounting methods and practices of the buyer and seller differ?

The area that causes most trouble between buyer and seller, and that is most often the cause for suspending merger negotiations, is inventories. The acquiring company views inventories as a banker might: "What could I recover from the liquidation of this inventory?" This is probably on the low side, and the selling company can point out that the buyer is, after all, acquiring a going business with all the potential inherent in it.

Obsolescence is a factor the selling company may not recognize until its CPA points out that some of the inventory may have little or no market value; on the other hand, the accountant may be able to point out some items of inventory that have been written off but would still have value to an acquirer. These points should be known to the seller before going into price negotiations.

In a recent merger the CPA was auditing the inventory of the seller, who manufactured aircraft flight instruments. The CPA came across a certain model gyroscope that was valued at its original price, although these items had been on the shelf for years. The accountant pointed out that the only aircraft that used this particular gyroscope had been out of production for some time, and the firm then agreed that it should write off this part of the inventory as obsolete. The CPA effectively strengthened the seller's negotiating position by eliminating this potential obstacle to the merger.

In order to detect changes in business practices and have answers ready for the acquirer's team, the CPA for the seller should make a trends analysis (of expenditures for various purposes, sales versus market penetration, sales versus sales expenses, R&D expenses, and so on).

The CPA as tax counsel must understand the advantages and disadvantages of the various forms of merger (discussed in the next chapter), and decide with the attorney and broker what forms may be acceptable, desirable or unacceptable. He can also assist the merger broker in the important aspect of determining what the savings resulting from the combined operation would be worth to the acquirer, and how much of this can probably be reflected in the selling price. He should also compute the effect of "pooling of interest" versus "purchase" accounting in terms of the price an acquirer will be willing to pay.

INVESTMENT ADVISOR

When the owners of the selling company elect to receive stock or other securities in the acquiring corporation, they are making a major investment decision. How can they be reasonably sure their newly acquired equities will not suddenly dive from twenty-five down to ten—that is, that their value will have somewhat better glide characteristics than a rock?

In most cases involving the merger of a closely-held company, the owners are so intent on price, selection of a buyer, making plans for turning over the reins and for their post-merger activities, that they neglect to scrutinize the value of what they are going to receive. Failure to analyze the value of securities proffered by an acquirer could prove to be very expensive; but accepting cash, without having looked at the merits of a buyer's securities, can be even more expensive.

Rate of growth, in terms of a diversified company's earnings per share, is not necessarily an indicator of continuing growth and stability. If growth in earnings per share is based on the acquisition of companies that had, prior to their merging, a higher comparative earnings per share, the parent corporation's growth may be due more to its skill in acquisitions and financing rather than its ability to generate profits on operations. At some point in time the value of the acquiring company's equities must depend on internally generated profits rather than, or in addition to, its skill in acquisitions and financing. The merger as an investment transaction must consider these factors. The investment advisor, therefore, may be a key member of the seller's merger team.

In selecting an investment counselor, the seller should seek a strong research capability and an ability to determine business values, such as the value to be obtained by the combination of the two firms (determining whether two plus two equals five or three). The advisor must obviously also have a professional understanding of the financial market and the buyer's strength and "credibility" in the market. (Two plus two equaled three in the merger of a paint manufacturer with an office equipment manufacturer; the stock of the merged companies immediately went down, because to the financial community there was no credible reason for the merger.) Another factor to look into is the *time* an investment counselor will give in providing the service that the seller seeks.

A recognized investment service or an investment banking firm can help analyze investment possibilities and provide counsel as an integral part of their normal operations. Some can provide not only factual information from which the seller can formulate his own decision, but will also make specific recommendations as to the best investment of the various possibilities available.

USE OF OTHER SPECIALISTS

Some of the other specialists whose services may be helpful in selling the closely-held company are the management consultant, the company's banker, the finder and the appraiser.

The primary function of a management consultant is no different in a merger transaction than in normal business operations: to strengthen a general or specific aspect of the business. The smaller company in particular can benefit from an experienced consultant, since he can provide the kind of expertise the closely-held firm may not be able to afford on its regular payroll. In selecting a consultant the company's chief executive should satisfy himself as to the capabilities, experience and qualifications of the firm and the individuals who will provide these services; and he should have a clearly stated arrangement as to what is to be accomplished, who will do the work, the fee arrangement and reports or other information to be provided. A service provided by some consulting firms is the merger consultant, who performs the analyses and forecasts and recommends steps to be taken to make the company more mergerable. This is on a regular fee basis rather than a contingent basis.

Because of pressures brought by their clients, some commercial banks have developed a capability for assisting in merger transactions. Primarily, banks are interested in not losing a client through the merger route; secondarily, they may be interested in providing as much service as a competitor bank. This service may run from informal counsel and serving as "finder" to full-fledged merger broker functions.

The much-discussed finder is another specialist whom the selling company may employ, knowingly or unknowingly. By reputation the finder is a rather dubious tipster whose sole activity is putting buyers and sellers in touch with each other, and taking a fee when a merger is effected. Figure 9–1 is a typical finder's letter, with names disguised. Finders may be respected professionals, however—bankers, lawyers, CPA's—who fundamentally earn a living practicing their basic occupation, but who occasionally spot the counterpart to a mergerable company. At this point the less professional of these professionals may succumb to the temptation to serve as a finder; and unless the seller is aware of this, he may find himself in an awkward position *vis-à-vis* the finder. One expert on the legalities of this possibility strongly recom-

Figure 9–1

Finder's Letter

September 19, 1970

William Q. Craven, Jr., President
Luminescence Incorporated
2400 Waddell Avenue
San Francisco, California

Dear Mr. Craven:

This is the era of easy and profitable mergers. With the right
match of companies, large gains can and often do result.

Would you consider opportunities in your industry — to buy or sell?

Bringing together qualified buyers and sellers is our business.

It is our only business and has been for ten years. Our reputation
for in-depth knowledge of companies and for matching them properly
has resulted in a long list of prominent clients.

Recently we handled transactions for Continental Cement, Crossitron
and Manco among twenty-one deals last year.

May we tell you about firms in your field? Then take a moment
to indicate your interest on the enclosed card.

We thank you . . . and we await your reply.

Very truly yours,

CORPORATE GROWTH ASSISTANCE

mends that the seller ask the individual who approaches him whether he is acting as a finder; and that he then document the situation as soon as he can, to the extent of backing this up with a confirming letter to the individual concerned.

BRINGING IN THE SPECIALISTS

Experienced acquisitioners, including lawyers and accountants, in major diversified companies note that lawyers and accountants can ruin deals faster than anyone else, and point to the necessity for selecting experienced individuals. Generally, too, these same professionals advise that initial discussions should *not* include lawyers and accountants; not so much because of their fees, but because the business merits of a deal need first to be explored by the heads of the two companies. The merger broker particularly can give good advice on when to bring aboard these and other specialists.

One suggestion in working with the various advisors: The chief executive should recognize that he might not get good *business* advice from the individual who specializes, whether he is a lawyer, CPA or banker. Planning, controls, market potential, product line expansion and organization and staffing are matters that normally are outside the expertise of these specialists. If the seller wants business advice, the merger broker is normally a better qualified source on these matters.

One further caution is advisable. Before the seller contacts a specialist, it is worth pondering, "Now if I ask my banker (for instance) who else might hear that I'm thinking of selling?" "If my CPA firm happens to do business with a competitor of mine, how might that affect the sale of my company?" "How tight-lipped is the lawyer I've been referred to?"

The answer to these qualms, as to the qualifications of the specialists he seeks, is the degree of their merger experience. In probably no other transaction in his career has the owner-manager stood to gain so much from the retention of experts, with pertinent experience.

CONCLUSION

The reasons for bringing in qualified specialists are these:

- Experience in varied merger transactions
- Expertise in tax, legal, accounting and other complexities of mergers
- Experience in negotiating and counseling under differing circumstances
- A specialist can work for the company without revealing the company's name
- Introduction of an outside viewpoint
- A specialist costs less than internal, permanent staffing for the same function
- Specialists save company executives' time

CHAPTER 10

MERGING: FORM OF THE TRANSACTION

The need for specialist assistance comes sharply into focus when the seller and the buyer begin to talk specific aspects of the transaction. It is in the final stages that a deal is most likely to break down, and the presence of experienced professionals who focus on the intent of the two parties may be the difference between seeing the merger through or seeing the meeting of the minds break up because of technicalities.

Effecting the deal requires knowledge of tax aspects because these usually determine the form of the merger; and it involves negotiations and hammering out of the contract that will bind both parties. Each of these is the subject of a broad field in itself, requiring constant study and scrutiny by professionals. While the businessman does not need the depth of knowledge possessed by his merger specialists, he should nevertheless have a working knowledge of these topics, in the same way that he must know at least the elements of business law, accounting and control.

"TAX-FREE" TRANSACTIONS

To this point "mergers" have been discussed as if the term means, simply and clearly, the sale of one company to another. This is adequate for general business usage. However, from a tax point of view, "merger" is a generic term covering six different tax-free transactions and various taxable transactions.

The Internal Revenue Code provides that, in general, when property is sold at a gain, the seller must pay taxes on the gain; and this is true in the sale of a business as it is with other property. There are six exceptions to this general rule in the case of business reorganizations, however, as defined in Section 368 of the 1954 Internal Revenue Code. Three of these exceptions (*A, B,* and *C*) pertain to the sale of one company to another (see Appendix B, pages 171–74). Each of these three (*A*—statutory merger; *B*—Stock for stock exchange; *C*—stock for assets exchange) has relative advantages and disadvantages to the seller and to the buyer. Each will be reviewed briefly for its impact on the deal proposed to (or by) the seller.

However, before considering the three forms of "tax-free" merger, it is necessary to look at what is meant by "tax-free." To the seller this term means that tax on gain resulting from the transaction is *deferred* until the stock received from the acquiring corporation is sold or otherwise disposed. It is possible, however, for the transaction to be truly *tax-free* where, for instance, the seller's estate planning provides for sale or transfer of equities held at the time of his death; in this case the gain realized, at the time of the transaction and subsequently, assumes a new tax base, as determined by its value at the time of his death. Often this precludes tax on the gain.

A "tax-free" transaction often involves the transfer of cash and other taxable benefits to the owners of the selling company; although most of the payment may be in equities of the buyer (and hence not taxed, because of the underlying assumption— and a very valid one—that the seller is merely exchanging his own business with its risks for the risks of another business, to the extent he receives equity in that business), the taxable remainder is subject to tax at the time of the merger.

As a general observation, the buying company will in most cases favor whatever type of merger most favors the on-going business interests. The seller will more likely be influenced by tax considerations.

One other observation should be noted: In a tax-free transaction, the provisions of the Internal Revenue Code must be strictly complied with or the entire transaction may be taxable to the seller. There are minor exceptions, and exceptions to these few exceptions; but the seller should expect to comply completely with the provisions or seek an advance ruling if even a slight deviation is contemplated.

Statutory Merger

A statutory merger is so named because it is carried out in accordance with the statutes of the state or states in which the buyer and seller are incorporated. In this kind of merger, or *A* reorganization, the shareholders of each corporation are required to approve the agreement of merger as approved by their boards of directors. Approval of one-half to two-thirds of the stockholders is required, as determined by statute. Dissenting stockholders in each corporation have a right to appraisal and to payment of fair value for their stock. The selling corporation goes out of existence as a separate corporate entity; normally the selling company liquidates and distributes the buyer's stock and "boot," if any, to its stockholders.

A difference between the statutory merger and type *B* and *C* mergers is that no closing documents, other than the agreement of merger, are required (although there will, of course, be a number of certificates, statements, and contractual documents exchanged between the two parties themselves).

Another principal difference in this type merger from the other two types lies in the options of payment available to the acquiring company. The acquirer can offer nonvoting common or nonvoting preferred stock, as well as voting stock; or a combination of these. In addition, he can offer cash or other property (debentures, bonds, government securities and so forth). This other property, called *boot,* is taxable (unless there is a capital loss instead of a capital gain). Where boot is part of the payment to the seller, the seller must maintain a "sufficient" proprietary interest in the acquiring company or be subject to taxation on the buyer's securities he receives as well as being taxed on the boot. What is "sufficient" proprietary interest is not specifically defined by IRS or the courts, although generally the stockholders of the selling corporation must receive stock interest in the acquiring company equal in value to at least 50 per cent of the value of their disappearing corporation.

To illustrate the effect of taxable boot in an essentially "tax-free" transaction, suppose that a stockholder previously paid $100 for a share in his selling corporation; and that he receives, in exchange for this share, the following from the acquiring corporation:

	Market Value
Two shares of buyer's common stock	$140
One share of buyer's nonvoting preferred	20
$15 cash	15
$25 in government securities	25
Total received	$200

In this case the stockholder has $100 capital gain; and he is subject to income tax on the boot ($15 cash and $25 government securities); but he receives the buyer's common and preferred stock "tax-free," since he retains proprietary interest to this extent in exchange for his interest in the selling corporation.

This flexibility in means of payment available to the acquiring corporation may help bridge the gap between the price asked by the seller and that offered by the buyer, and is a prime reason for using an *A* or statutory merger.

Stock-for-Stock Exchange

In this type of acquisition (*B* reorganization), the acquiring corporation must exchange its *voting* stock (common or preferred, or a combination) for the stock of the selling company; and the acquirer must, immediately after the exchange, have *control* of the acquired company. (Most sellers of closely-held companies will effect the sale completely in one transaction. Between larger companies, however, an acquirer may receive less than controlling interest in one transaction, and subsequently acquire additional stock in another transaction; the exchange does not become tax-free until the acquirer has control of the acquired company.) *The stockholders of the selling corporation may not receive cash or other forms of payment, either directly or indirectly,* from the acquirer, other than voting stock. In fact if some stockholders receive payment or even partial payment in a form other than voting stock, the transaction may revert to a taxable reorganization, subjecting *all* selling stockholders to taxation on gains. A selling company may distribute its own cash prior to the transaction, however, without thus affecting the nontaxable nature of the merger.

In order to meet the requirement for control, it is necessary that the acquiring company receive 80 per cent or more of *each* class of stock of the seller; if he receives less than 80 per cent of any class (or if, with stock previously acquired, the total is less than 80 per cent), the transaction will be taxable.

In a stock-for-stock exchange the acquired corporation retains its identity (as a subsidiary) and also retains its obligations for unrecorded liabilities. The selling stockholders who participate in the transaction merely exchange their stock for the voting stock of the purchasing corporation (or of its parent corporation, but not for a combination of both). The transaction is between the buyer and the stockholders of the selling corporation, bypassing the selling corporation itself. Because of this procedure, this is often the simplest of the several types of reorganization. Outstanding minority interests may be a problem, however; and if the minority interest is sufficient to prevent the acquiring company from getting as much as 80 per cent of each class of stock, the *B* transaction may prove undesirable for either the buyer or selling stockholders. (In this connection it should be noted that there are no appraisal rights for dissenting stockholders.) Union contracts, pension plans, contracts to perform services, and leases usually carry over after the transaction.

Assets-for-Stock Exchange

This type acquisition (*C* reorganization) has both similarities and differences compared to each of the transactions just discussed. Approval of the selling company's stockholders is required (but not that of the buyer's stockholders), although the transaction is between the buying and selling corporations rather than between the buying company and the seller's stockholders. The selling corporation is lost to the buyer; however, the buyer does not assume responsibility for unrecorded or contingent liabilities, such as Federal income taxes, *unless specifically stipulated.* (This point should be noted by the selling company's merger attorney.) Dissenting stockholders of the selling company have appraisal rights, but generally no outstanding minority interests will remain after the transaction; this may be a significant business consideration to the acquiring company.

In an assets-for-stock deal, "substantially all" the assets—generally considered 90 per cent or more of the seller's properties (net of liabilities)—are exchanged "solely" for voting stock of the buyer. There are exceptions to "solely for voting stock," however, in contrast to the very strict ruling under a *B* (stock-for-stock) transaction. For instance, where the acquirer gives voting stock for more than 80 per cent of the value of the seller's assets, he may be permitted to use cash or property for the balance. However, the effect of outstanding liabilities may nullify this possibility if the total of these liabilities and cash or other property exceeds 20 per cent of the value of the selling company's gross assets.

Summary Comments

From the preceding discussion regarding each of the three forms of reorganization, it is apparent that each has its advantages and disadvantages to the two parties to a merger; in actual practice, circumstances will probably dictate which is the most practical to use.

A precaution that should be observed is the case where employee-stockholders of the selling corporation want employment contracts for stock options from the acquiring company. The two companies should determine, through reasonableness of remuneration and reference to expert tax advisors, that the payments under the contracts cannot be construed as payment under the merger transaction, thereby upsetting the nontaxable nature of the deal.

Tax loss carry-overs would also affect the price, since the acquiring company would benefit from them and could therefore pay more. In this event the selling stockholders should evaluate whether a taxable transaction, in which they would receive payment entirely in cash and other property, and pay tax at the capital gains rate, might be more to their advantage than a nontaxable transaction.

TAXABLE TRANSACTIONS

The owners of a selling company can elect to have a taxable transaction, and often do. This has the advantage of winding up the merger in a clean way by paying the 25 per cent capital gains tax. If the business investment originally was $100,000 and the company is subsequently sold for $1.1 million, the owners would receive their original investment and 75 per cent of the $1 million gain, or a total of $850,000. They may be able to reduce the effects of taxation by spreading the payments over a number of years, so long as not more than 30 per cent is received in the year of the transaction.

If there is a loss rather than a capital gain, the owner may be able to use the loss carry-back to obtain a refund of income taxes previously paid. The acquiring company may also want this tax loss, however, and would therefore be inclined to propose a taxable deal to the seller. Each side, in other words, wants to use a tax loss to its own advantage, and will seek that form of transaction which is most favorable to itself. The seller may receive a lower price if he is the one who obtains the benefit of the tax loss; but the net after-tax effect may be more to his advantage than a higher price on which he pays full taxes, or on a "tax-free" transaction.

In a sale of assets, the sellers may be subject to double taxation—once through tax on the selling corporation, and a second time on surrender of their shares in it—

unless certain precautions are followed. The Internal Revenue Code provides that the sellers can adopt a plan of liquidation (twelve-month or one-month plan) and carry it out; they will then be subject to taxable gain on only their individual shares. This precludes tax on the company, inasmuch as the corporation is effectively acting as an agent for the stockholders in putting its assets in their hands.

Normally the owners of a selling company will elect, as a body, whether to have either a taxable or a tax-free transaction. It may be, however, that some stockholders want a taxable deal and others want a tax-free deal. This can be accomplished by the use of a statutory merger, in which some will receive cash and others will receive voting or nonvoting stock. The former will be taxed, and the latter will achieve their tax-free objectives. An advance ruling from the IRS should be obtained where a mixed result is sought.

CONCLUSION

This discussion on taxable and tax-free transactions is general, and there are numerous exceptions and exceptions to exceptions that may apply in particular cases and require expert tax and legal counsel. The seller may be inclined to elect the taxable route, for the simple reason that it is a quick, clean way of completing the transaction and because he is not himself sufficiently skilled in these matters. The advantages of a tax-free transaction, however, are worth exploring with an experienced specialist. Even if the specialist recommends a taxable form of merger, the seller will know he has received the fullest value from the transaction, and this mental satisfaction will remain with him for a long time.

CHAPTER **11**

MERGING NEGOTIATIONS

Negotiation is defined as a mutual discussion and arrangement of the terms of a transaction or agreement. It is an attempt to find a formula for the merger that will maximize the interests of both buyer and seller. Professional negotiation is neither haggling, "horse trading" nor winning an argument; instead, it is a process of communication that reflects a balance of both parties' short-run and long-run interests.

In negotiations, as in other aspects of the merger activity, experience pays off. A buyer may have been through negotiations in as many as thirty or forty acquisitions, whereas most sellers have never before been involved in merger negotiations. In most cases a buyer will try to be fair to a seller; but obviously he would be derelict in his responsibilities to his own company if in negotiations he tried to look after the best interests of both his own company and that of the seller.

There are ways in which the seller can prepare for negotiations, however, including ways of compensating for his personal inexperience. This chapter will assist him in these preparations.

SEQUENCE OF NEGOTIATIONS

Negotiations start with informal discussions between the buyer and the seller. The buyer may be an officer other than the acquiring company's chief executive. For example, in one acquisition-minded company, the area vice-president makes the initial contact, getting an idea of the fit and the asking price. If necessary, he then brings in the company's executive vice-president to complete the transaction. If this individual does not close the deal, the president conducts negotiations.

The selling company's chief executive may conduct negotiations himself, or he may have an experienced merger broker assume this responsibility. For several reasons the selling company's chief executive may remove himself from negotiations. First a raspy personality, on either side, can turn off a transaction that would have been good business for both. Another reason is that the seller can put himself in the "good-guy" role; that is, let his representative press his opposite number on technical points, and then himself graciously take off some of this pressure in the interests of achieving the merger's overall objectives. In addition, negotiating can require a considerable amount of time.

The Business Agreement

Basically there are two agreements that are negotiated: the business agreement and the legal agreement.

As mentioned previously, it is important that the two principal negotiators have an opportunity to explore the business benefits and "fit" of the proposed merger before getting involved in legal and technical discussions. They should also explore objectives, areas of mutual interests, complementary strengths and other bases for the merger before getting involved in price discussions, which might cloud the basic purposes of the contemplated merger.

Price. After the benefits of a merger become apparent, the acquirer will get around to the prime question: "What are you asking for it?" The seller should probably hedge in replying, since price relates to terms and conditions of the sale and these probably have not yet been discussed. He might reply by pointing out the company's profit trend and by citing an applicable price/earnings multiple. *He should be well-prepared at this time, based on his CPA's analyses and his own preparation, to present his company in its best light, and to answer adverse points the buyer may bring up.* He should also know the objectives of the buyer, the savings or other advantages to him and the form of payment the acquirer may have in mind.

Terms and Form. The discussion may then get into terms and form; the pros and cons of 100 per cent payment at closing or payment up to 30 per cent at closing and the balance over a period of several years, and whether the seller wants cash or will accept the equity or securities of the buyer. If equity in the buyer's company is considered, the seller should inquire early in discussions as to the buyer's outstanding stock options, internal versus acquisition-generated profit and the expected effect of regulatory actions on continued growth. The purpose of bringing these points out early is to let the buyer realize negotiations will be a two-way street pertaining to the strengths and weaknesses of both companies, rather than just those of the selling company.

This leads to whether the merger is a sale of stock or assets, or is a statutory merger. The buyer may be restricted as to the method of payment available to him, and this may in turn affect the terms, form and price of the proposed transaction.

Negotiations on price and terms may require several go-rounds before the gap between buyer and seller narrows to agreement. If fundamental benefits of the merger are obvious but price remains a barrier, either party may suggest an earn-out. The seller should carefully evaluate the effects of an earn-out proposal, however.

Additional Items of Negotiation. The company's name and location may be important to the owners of a closely-held company, as well as such matters as employee benefits, brand names, compensation, and fringe benefits for employees. The seller may be interested in where he will fit into a management role after the merger, but he should not let this separate matter adversely affect his responsibility for negotiating the best possible sale of the company.

The seller should recognize, of course, that the acquiring company is likely to make substantial changes—in controls, possible relocation of facilities and realignment with their other operations—and that business as usual is unlikely. For a seller to try to negotiate business as usual is inconsistent with negotiating the best price and terms.

Reaching an Agreement. When they have reached an agreement, the two principals should *write out* the areas of understanding so that each party will have a clear record of the agreement. Failure to do this may cause delays and require additional negotiations.

Where a broker conducts negotiations for the seller, he may receive an offer from the buyer and transmit it to the seller ("I am authorized to present this offer to you"), and in turn transmit the seller's response to the buyer. Being in close touch with both parties, he is often in a position to evaluate what the other party will or will not accept, and can simplify negotiations by precluding offers or counter offers that do not lead constructively to a final agreement.

Legal Negotiations

After the fundamentals of the business transaction are agreed to, the second negotiation, the legal framework, begins.

Where possible, it is desirable to get the two lawyers and their principals *in the same room* to reduce the businessmen's agreement to a definitive legal document. In this way there is no wondering, "What did he mean by that?" Each individual can ask the other party's intent directly. Frequently the attorneys can resolve a question with, "How about saying it like this?" And the other may respond, "Sure, that's all right."

Where the lawyers deal at arm's length, however, negotiations can stretch out for months after the basic business agreement. Pride of authorship and questions of intent and phrasing can touch off round after round of correspondence. In one case the Purchase and Sell document was drafted by the buyer's attorney in accordance with the principals' agreement, and was delivered to the seller's lawyer. The latter did not even look at the document for ten days; subsequently he began to question wording and, eventually, the substance of the agreement. It was more than sixty days until he finally concurred in the document, essentially as originally drafted. In another situation more than four months elapsed because of similar delays.

In a contrasting transaction, the buyer's attorney (who resided in a different city) arrived at the seller's office at 9:00 A.M. for the purpose of drafting the document in conference with the principals and the seller's attorney. After four hours a final copy of the legal document was dictated, and by the end of the day a typed copy was in the hands of each participant.

The Purchase and Sell agreement is usually signed when finalized, although signing may be delayed until closing.

Seller Warranties. A seller can reasonably expect to sign a warranty that he will continue to conduct his business in the same way from the date of agreement to the closing of the transaction; that is, that he will not incur any unusual liabilities or obligations, will not declare dividends other than those regularly distributed, will not make capital expenditures without buyer approval, and so on. He will, in other words, continue the business as if it were not being sold.

These warranties as to business *practices* are quite different from warranties as to the business' *condition*—the condition of contract liabilities, contingencies, patents, real estate, inventory, and so forth. On some of these items, such as patents and royalties, he may expect to bear a continuing responsibility per his warranty to the buyer. But he should seek to minimize the warranties that bind him. He can inform the buyer that he has nothing to hide and that the buyer may investigate to whatever extent he wants, but that once the deal is closed, it stays closed. From that point on, the buyer assumes responsibility. The seller will probably have to make some concessions on this point, depending on the buyer's opportunity to investigate thoroughly. But the fewer warranties he makes, the less continuing liability the seller will bear.

Part of the purchase price may be placed in escrow with a plan for payment, if this is negotiated. In a typical example where a company sold for $5.5 million, $500,000 was set up in escrow for liabilities, particularly taxes, to be released to the seller during a five-year period. (Interest on earnings of the escrowed amount normally goes to the seller, although this is a matter of negotiation.)

An effective guarantee against bad debts is making accounts receivable part of the payment to the seller, and this is often used in merger agreements. Matters of incorporation and patents are often warranted by lawyers' documented opinions rather than seller warranties.

Buyer Warranties. The buyer will also make certain warranties, although these are usually much fewer than those of the seller. One authority states: "A sophisticated seller is interested in a buyer's representations and warranties, particularly if the seller is taking any kind of buyer's paper [common stock, preferred stock, convertible debentures, notes] in payment for the transaction. Sometimes a seller does not insist properly on representations and warranties from a buyer."[1]

In addition to matters regarding his authority to conduct the transaction and representations as to his paper, the buyer will warrant such items as a term pay-out, conditions of an earn-out and terms of lease if applicable.

The buyer's attorneys furnish documented opinions as to matters of incorporation, similar to the opinions provided by the seller's attorneys.

Rights of Inspection. The seller must ordinarily grant rights of inspection to the buyer. He can expect that terms of inventory, particularly, will receive thorough scrutiny from the buyer's accountants and production personnel. The reasonableness of this is indicated by a potential acquisition in which Litton's investigators discovered that substantial items of inventory were missing, and that neither the sellers nor their auditors had been aware of this fact.

Escape Conditions. The parties generally will have no difficulty in reaching agreement on their rights to discontinue merger proceedings. This includes such matters as discontinuing if an insufficient number of stockholders in either corporation concur in the proposed merger; discontinuing in the event of government antitrust action; discontinuing if conditions of the seller's company as audited are found to differ from the practices or conditions that are the basis for negotiations, or discontinuing the merger proceedings if the condition of the buyer or his securities at closing differs substantially from his warranty at the time of negotiations. Any misrepresentation on the part of either buyer or seller is grounds for not closing.

PREPARING for NEGOTIATIONS

In most cases sellers have not prepared for negotiations to the extent that they have analyzed the acquirer's strengths and weaknesses relative to their own, or have gone through dry-run negotiations. It is particularly important that the seller have a good understanding of the buyer's objectives so that he will know if they are compatible with his own.

Know the Acquirer

From published information, from principals of companies who have previously sold to the same buyer and from bankers and other sources having knowledge of the inner workings of a buying company, the seller can usually find out significant information that will be important in negotiations. As a minimum he should know the buyer's areas of strength and weakness; the buyer's reasons for wanting to acquire the seller's company; and the individual who will probably conduct negotiations for the buyer, and something about him as an individual.

---

[1] Charles A. Scharf, *The Business of Acquisitions and Mergers*, G. Scott Hutchison, ed. (New York: President's Publishing House, 1968), p. 295.

The seller should also find out the acquiring company's behavior patterns on previous negotiations ("Let's go out and discuss it over a couple of martinis"—"Come on up to our headquarters and I'll show you how much we can do for your company"), his extent of "give," the reasons given for proposing certain forms of transaction and for not proposing others and his pattern of terms and post-merger operations (whether he lives up to pre-merger promises).

The buyer's plans and the merits of his stocks and securities (assuming the buyer may offer paper, as most acquisition-minded companies do) are particularly significant. The seller might ask his investment advisor for an appraisal, and ask him to be available when negotiations come to this key point.

There will be certain areas in which the acquirer will have little room to negotiate, such as a price that would adversely effect his earnings per share and continuing a modus operandi that would preclude his making the profit expected.

Know Himself

The seller-negotiator must have clearly in mind his own company's strengths, weaknesses and objectives, and must be prepared, as mentioned in Chapter 7, to answer questions pertaining to weak areas. He should have arrived at an objective appraisal of why his company is valuable to the potential merger partner, the minimum and likely prices he will accept and the maximum he can reasonably expect. Documented information on the company's history, earnings records and forecasts, ownership, and so forth should be on hand for discussion and presentation to the acquirer.

Role-Play

At least twice prior to actual negotiations, the seller should go through a role-playing exercise, with either another solid businessman or competent merger broker (not with a lawyer, banker or CPA). Role-playing serves to clarify issues and the process of negotiations, and gives the seller a feel for the buyer's position. He will also find where not to "give," and how soon to give in other areas. The seller should assume his own role and that of the buyer in separate sessions.

SUGGESTIONS on NEGOTIATIONS

On some points, of course, there will be no objections by either party. These can be handled quickly and need no further discussion. But assuming that the seller will start from an optimistic price and the buyer will start from a pessimistic one, there will probably be a great deal on which the two parties disagree.

Record the Issues

It may be helpful to come back to points requiring further negotiation by setting them down in an organized way, such as:

ISSUE: INVENTORY

Buyer's Position	*Seller's Position*
Estimates of completion are overstated, considering negative **PERT** slack (schedule slippage) on key items; and the method of associating progress with costs incurred is subjective and optimistic relative to the buyer's own practices. Since these are fixed-price contracts, indicated overruns should offset the merger price.	Work in process on government contracts *A, D* and *G* are accurately stated in terms of percentage completion and expenditures incurred or committed relative to this progress. The company's record of technical, cost and schedule performance is at least as good as the average for the industry.

The seller should organize the issues in this way, although the buyer may also do this. A clear statement of the issues will enable negotiations to proceed in a relatively clear, logical manner.

Reduce Agreements to Writing

Points of agreement should be reduced to writing as soon as there is an opportunity to do so, for instance, during a recess or other break. These will not, of course, be expressed in full legal terminology, but should reflect the agreement in principle; and each party should indicate verbal concurrence to what is written. This will help prevent "buyer's remorse" or, more likely, "seller's remorse," as often happens where the seller's emotional attachment for his company begins to affect his effectiveness as a negotiator. More importantly, this will ensure that agreements resulting from negotiations correctly reflect the parties' intent.

Set a Time for the Next Session

It is not unusual for negotiations to break off on a point of disagreement, particularly when the two sides have reached key points. Each party should recognize that the closer they come to satisfactorily concluding the transaction, the more likely it is that obstacles will arise that could prevent the merger. It is important that a time and place be set for continuing negotiations; otherwise the normal press of business can interfere with merger proceedings. Timing is particularly important to a buyer because he may be under greater pressure as a result of regulatory aspects, stock market fluctuations, other prospective acquisitions and the administrative costs of carrying out an acquisition program. For the seller, timing is important because the value of his company may fluctuate due to supply and demand.

Remember the Human Factor

The intensity and duration of negotiations is a drain on an individual. Occasional humor, taking a nap if there is an opportunity, getting a change of scenery from the ne-

gotiating table and discussion of other topics during breaks can do wonders for a-
chieving agreement in negotiations. Saving face can be important; and each side
should be perceptive to the needs of the other where an "out" is vital to agreement on
a significant point.

NEGOTIATIONS in LARGER COMPANIES

Where the selling company itself is a sizable organization, no one individual is likely
to have sufficient knowledge of law, accounting, manufacturing, sales, engineering
and general business to conduct negotiations himself. The acquiring organization may
therefore have a team involved in negotiations, and the procedure becomes more
formal—such as using an agenda—than in the discussions involving a smaller company. The seller needs advisors on call or actually present who correspond to the
buyer's counterparts, so that he is not automatically outnumbered and outgunned.

Organizing for Negotiations

The selling company's senior negotiator—chief executive or merger broker—should
clearly define the responsibilities of each team member. Negotiation is not a discussion between individual specialists; and specialists, such as lawyers and accountants,
should be aware of their responsibilities to advance the overall objective of the negotiation although they may have to "give" on some points pertaining to their own pro-
fession. The success of a negotiating team depends on its ability to advance collec-
tively toward achieving the sellers' overall objectives.

Qualifications and Attributes

The senior negotiator must be acquainted with the background of events leading to
negotiations, must be able to plan negotiations and carry them out effectively and
must be capable of controlling the team members. He must be a strong individual;
but this strength derives from firmness, knowledge, ability to communicate effectively, tact and judgment, rather than from being loud, stubborn or impatient. The more
he prepares himself to take part in the entire merger transaction, and the more prepared he is in specialists' areas, the greater his contribution during negotiations.

Some of the attributes of a good negotiator, in addition to knowledge and attitude are:

- He must be a clear and rapid thinker, for obvious reasons.

- He must be able to analyze alternative points of view and effects of proposals,
  and must be objective enough to consider other persons' ideas.

- He must be tactful, have a sense of humor and be impersonal enough to control his temper even when provoked. (Provoking the other side's negotiators is a technique sometimes used in negotiation to draw attention to minor points at the expense of more significant ones.)

- He must be patient enough to hear out specialists, but know when to step in if necessary to avoid getting bogged down in detail.

- He must be able to communicate well. This extends to rapport, but means particularly the ability to convey ideas and concepts to others and to clarify exchanges between others.

As the captain of his team, the senior negotiator should determine the extent of these capabilities in each of the other members of the team prior to negotiations. Role-playing will give him an opportunity to observe the specialists in action. In addition, he needs to know each team member's strengths and weaknesses, technical competence and experience in merger negotiations.

Strategy and Tactics

The advantage in negotiations often lies with the side that has the initiative, beginning with setting the agenda. For instance, if an agenda focuses initially on weaknesses of the selling organization, it may be difficult to recoup this initial effect on price/earnings multiple during subsequent discussions.

If the seller is instrumental in setting the agenda, he should align his main guns at the start and intersperse strengths of lesser importance along with weaknesses later on the agenda, since the initial impression makes considerably more impact than points later discussed.

In his preparation the senior negotiator, with respective specialists, should determine a minimum, maximum and likely position for each point on the agenda. This does not mean that, in total, any one position cannot be accepted at below minimum, when considering the overall effect; but it will enable him to argue more effectively on each point in order to optimize the overall results.

The first go-through of the agenda may be just a feeling-out of each side by the other, without substantial give and take; and this first pass may be undertaken primarily by each of the two senior negotiators. Subsequent passes will probably involve specialists and detailed discussions are likely to develop—again, mainly for clarifying points of negotiation rather than for arriving at agreement. From this will emerge major and minor points and relative positions, which then will involve standing firm on some points and yielding on others. If the discussion can be so guided by the seller, he might take up items in such a sequence that he will yield on one point in order to gain the next point, which is more important to him.

An individual, perhaps the legal representative, should be designated to reduce agreements to writing.

CONCLUSION

Negotiation is the culmination of years of building a company to the point where it is valuable as a merger partner. Successful negotiation is primarily a result of keeping in mind the objectives to be achieved during the discussions and, secondly, being well-prepared for negotiations.

The seller should recognize that the closer the two parties come to agreement, the stronger the factors will emerge—some very trivial—that could keep them apart. At this time no problem is too small to merit the full attention of the selling and the acquiring chief executives.

The words of the managing director of a major consulting company might be worth bearing in mind: "The saddest deals are those that didn't go through, but should have."

CHAPTER 12

MERGED! THE CONTRACT AND CLOSING

The owners of the selling company will, of course, have competent legal counsel who will put the two parties' Purchase and Sell agreement in proper legal form. The resulting contract is basically the agreement as to what is to be bought and sold (stock, cash or securities from the purchaser, in exchange for stock or assets of the seller), and includes warranties on all representations (such as the selling company's contractual obligations, and financial condition of the company as of the agreement and as of closing).

The closing, which is the transaction completing the merger, requires affidavits and certificates in addition to the terms of the contract. These are usually obtained during an interim period after the contractual agreement is effected. At closing, the buyer and the seller exchange each of the items required in fulfillment of their contractual obligations.

This chapter exposes the businessman-seller to the terms and conditions normally included in the contract and closing documents. It does not therefore go into depth in the legal aspects of these two formalities.[1]

THE CONTRACT

"The" contract is frequently several contracts. The major one of these documents is the Purchase and Sell agreement, which binds both parties to the merger. In addition, there may be a noncompete agreement, an employment contract, a lease agreement and other agreements. These may be exhibits appended to the Purchase and Sell

---

[1] For a more detailed discussion of legal aspects of a merger, see Charles A. Scharf, *Techniques for Buying, Selling and Merging Businesses* (Englewood Cliffs, N.J.: Prentice-Hall, 1964).

agreement or they may be completely separate from it, but each is a signed contractual document that stands on its own as a legal instrument.

The importance of the Purchase and Sell agreement is indicated by this quote: "... Because of the numerous possible pitfalls in buying another's business, the buyer should approach the acquisition as though the entire purchase price could be sacrificed as a result of some oversight."[2] The seller should be no less careful regarding *his* interests.

The Purchase and Sell agreement serves other purposes in addition to being a legal document. It is also a reference for questions that may subsequently arise, and it serves as a framework on which to hang all the other documents that are part of the merger procedures. Thus exhibits regarding patents, profit-sharing plan, and so on may be appended to the Purchase and Sell agreement.

The contents of an agreement covering the sale of his stock by a single owner-executive, a relatively simple transaction, are shown below. The headings are typical and from a merger contract in which the owner sold his company for more than $1 million cash, and the buyer was particularly interested in obtaining the seller's patents.

STOCK PURCHASE AGREEMENT

1. *Preliminary Definitions.* This clause defines the closing date, the selling company and individual, the purchasing corporation and other pertinent matters (such as "escrow agreement, " "consultation agreement") referred to in the contract.

2.   *Sale and Purchase of Shares; Payment of Purchase Price.*

(a) In this paragraph the seller agrees to sell to the purchaser, at the closing, all the shares of capital stock that he owns, together with patents; and the purchaser agrees to purchase these for an aggregate purchase price (specified), to be paid as set forth in paragraph 2b, below.

(b) *Purchase Price; Escrow.* The seller agrees to sell to purchaser all the shares of capital stock, subject to all terms and conditions of the contract, of his company (or companies, where one may be a manufacturing company and another a sales organization, for instance), for the specified purchase price. Seller agrees to sell to purchaser the patent or patents for a purchase price. Also, seller agrees to enter into a consultation agreement. The purchaser agrees to deliver to seller a certified or cashier's check in an amount specified (this may be all or part of the purchase price); some of the purchase price may be delivered to a bank in escrow if a fund is to be established for this purpose.

3.   *Closing.*   This clause states the time and place of closing, the documents to be delivered to the purchaser and the documents to be delivered to the seller. These documents and the closing event are discussed in more detail under "Closing" in this chapter (pages 144–46).

---

[2] *Ibid.,* p. 168.

4. *Representations and Warranties of the Seller.* An indication of the extent of representations and warranties asked for by the buyer is the fact that one of the leading diversified companies used to prepare a two-page document covering the contractual arrangement. The same company now typically prepares a seventy-five-page document, including exhibits. The seller's representations frequently will constitute one-third or one-half of the body of the agreement. As discussed in previous chapters, this section is vital to the interests of both the buyer and seller.

Some of the representations, warranties and agreements of the seller are these:

(a) That the selling company is organized and in good standing in the state in which incorporated, and that it has full corporate power and authority to conduct its business; whether the company has any subsidiaries; and that it is (or is not) a partner or joint venturer with any other entity or person.

(b) The total stock and par value of seller's capital stock; and the number of shares issued and outstanding.

(c) Seller is the owner of the outstanding capital stock, free and clear of liens, encumbrances, and so forth and has the right and authority to transfer and deliver these shares as provided in the contract.

(d) A listing of the present officers and directors of the selling company.

(e) Seller has delivered to buyer copies of statements of the financial condition as of (about the time of negotiations), with corresponding statements for prior years, along with CPA's notes and statements. Exhibit 1 usually includes these representations and warranties.

(f) All of the accounts and notes receivable of the seller are valid and collectible at face value.

(g) Description of real and personal properties owned by the seller and the title to them. These and their tax status would in many cases be detailed in an exhibit, as would insurance coverage on these items.

(h) Except as specifically stated otherwise, the company has no liabilities, commitments or obligations of *any* nature—whether absolute, accrued, contingent or otherwise, and whether due or to become due. Also, the seller knows of no basis for the assertion of liabilities, commitments, and so on against it.

(i) As of the date of signing the contract, the seller is not a party to any outstanding contract or other commitment involving more than (e.g.) $5,000 or having a duration of more than (e.g.) two months, except as otherwise specifically stated.

(j) The seller has no knowledge or basis for believing that any of his customers will terminate a material portion of their normal purchases from the company.

(k) The seller warrants the status of any agreement or understanding made other than in the usual course of his business; any employment contract not terminable, without cost or liability, on thirty days' notice; any contract or agreement with labor unions or other collective-bargaining groups; officer or employee bonus plans, profit-sharing plans, retirement arrangements, hospitalization plan, group insurance, pension or welfare plans; and any lease or agreement regarding use of land.

(l) No event that could materially and adversely affect the business prospects or conditions of the company has taken place which is not reflected in financial statements at the time of negotiations.

(m) Since the time of negotiations there has not been any change in the condition of the business except changes in the ordinary course of business, none of which are materially adverse; any damage or loss materially affecting the business or prospects; any declaration or setting aside of payment for dividends, or any purchase or acquisition of stock or securities issued by seller; any increase in compensation payable to seller's officers or employees, which in the aggregate exceed (e.g.) $1,000.

(n) Federal and state income tax status, and liability of the seller for any taxes subsequently discovered to be due.

(o) To the best of seller's knowledge, his company has performed all obligations, and he has no knowledge of threatened or actual material returns of goods or products, or other claims or causes of action.

(p) The seller will continue in force the kinds and amounts of insurance listed on an exhibit.

(q) Regarding patents (listed and described on an exhibit), there is no known infringement by others; and (for the seller's protection) he makes no representation or warranty that other proprietary subject matter is patentable.

(r) No litigation, proceeding or investigation is pending or threatened that would result in a material, adverse change in the selling company's prospects or conditions; and the seller knows of no basis for any such actions.

(s) Location of the company's facilities; and lease or other arrangement by the seller to the buyer.

(t) No statement made by the seller in connection with the contemplated transaction contains or will contain an untrue statement of a material fact, or omits to state a material fact.

5. *Representations and Warranties of the Purchaser.* In modest simplicity, the purchaser may warrant that he is a corporation organized and in good standing under the laws of the state of incorporation, and is authorized to do business in the seller's state; and that the merger agreements have been duly approved and authorized by the corporation, usually by the directors. In a stock-for-stock exchange, or in a long-term payout, however, the seller's warranties may be as encompassing as those of the buyer.

6. *Conditions of the Closing.*

(a) Any obligations of the purchaser are conditional upon the seller's having performed each of his obligations, and representations and warranties of the seller shall be true as of the closing.

(b) Obligations of the seller are conditional upon the purchaser's having performed each of his obligations, and representations and warranties of the purchaser shall be true as of the date of closing.

(c) The seller will try to retain the services of employees and preserve the goodwill of suppliers, customers and others for the benefit of the purchaser.

(d) Agreement by the seller to lease real property, at a stated rental and period of time, to the purchaser.

(e) The buyer's representations and warranties shall be true as of the date of closing.

7. *Access to Information and Documents; Confidential Information.* Here the seller agrees, pending consummation of the transaction, to give the purchaser full access to all books, contracts, properties, and so forth. Purchaser in turn agrees that he will not reveal any of the confidential information or documents to anyone other than his own officers, employees and representatives.

Anticipating the possibility that the transaction may not be consummated, purchaser agrees to return written information to the seller, and agrees that he will not reveal any information obtained during negotiations; also, he will not for (e.g.) three years use such information to manufacture, license, and so forth, products competitive with those of the seller.

8. *Survival of Representations and Warranties: Indemnification.* This clause provides that representations and warranties made by the seller and by the purchaser shall survive the closing.

(In the actual case involving this model contract, the seller agreed to have nearly 25 per cent of the purchase price put in escrow, payable to himself over a three-year period, pending any defect in warranties as to his patented device. In two years following the sale, $9,000 in penalties was assessed against the escrow fund because of defective finished goods that had to be scrapped.)

9. *Brokerage and Accountants' Fees.*

(a) The seller represents and warrants that all negotiations relative to the agreement have been carried on by them and their counsel without any other persons, except that the seller has agreed by separate agreement to compensate the (named) merger broker for his services in connection with consummation of this agreement. The seller holds purchaser harmless against any brokerage commissions from the (named) merger broker, and from commissions and so on arising from the transaction.

(b) The purchaser warrants that all negotiations for it have been carried on by his officers, employees and counsel; and that he holds the seller harmless against any claims for brokerage commission, compensation, and so on.

10. *Miscellaneous.* This clause pertains to changes and modifications; notices to be given by and to each party and their counsel; legal fees in the event of litigation; binding nature of the agreement on assigns, heirs, and so forth. The purchaser may assign or transfer his interest in the agreement to a wholly-owned subsidiary, and

may request at closing that the seller deliver and assign shares of his company's capital stock to this subsidiary. Finally, the agreement provides which state's laws will govern, where the two companies are incorporated in different states.

Schedule of Exhibits.   The exhibits referred to in the basic contract may include such items as these:

> Consultation agreement
> Escrow agreement
> Patents and patent applications
> Material contracts, profit-sharing plan, labor contracts, and so forth
> Officers and directors of the selling company
> Schedule of equipment and inventory (this may run from 60 to 100 pages)
> Schedule of insurance
> Preliminary title report
> Financial statements

This, then, is the nature of the document that determines the merger deal.

THE CLOSING

The closing can sometimes be accomplished at the same time that the contract is signed. Usually, however, additional affidavits and certificates are required and the date of closing is fixed for a later date, some thirty to sixty days after signing the Purchase and Sell agreement. (The signing of the Purchase and Sell agreement itself may be delayed until closing.)

A closing memorandum, prepared by legal counsel for the buyer and the seller, outlines what will happen at the closing. The introductory material in the memorandum cites the Purchase and Sell agreement between the buyer and seller, and the specific time and place of the closing. It also states who, by name, will be present. This normally includes the sellers (in a closely-held company); their merger broker; sellers' legal counsel, and patent counsel where patents may be a key factor; a bank representative, where an escrow agreement is part of the transaction; the buyer's representatives; and buyer's legal counsel. The memorandum then lists the actions taken prior to closing, actions to be taken at closing and those to be taken following the closing.

The actions *prior to closing* can often be stated very briefly: "The Stock Purchase Agreement was executed by Mr. Seller" and "The Board of Directors of the buyer met on (date) and authorized their officers to execute the Stock Purchase Agreement. Shortly afterwards Mr. President executed the agreement on behalf of buyer."

The closing memorandum then details what each of the parties will deliver *at closing.* In a sale of stock for cash, by a single owner, the seller or his lawyer would deliver items similar to these to the buyer:

- Certificates representing all of the issued and outstanding shares, endorsed in blank for transfer, with signatures guaranteed by a bank

- The seller's stock book

- Copies of an escrow agreement

- Copies of the consultation agreement

- Opinion of seller's counsel that the selling corporation is duly organized and licensed to carry on its business; the number of authorized and issued shares of stock and the par value; that the seller has the right and power to execute the various agreements and the right and power to convey title to his shares of stock; that the stock to be sold is free of encumbrances; that counsel does not know of any litigation that would adversely change the condition of seller's company; and opinion of seller's patent counsel that seller owns the patents and that at closing all of his interest and title in patents will transfer to the purchaser

- Resignations of officers of the selling company

- A certificate of selling company's secretary as to ownership of the company's issued and outstanding shares of capital stock

- Minute books of the company, with articles of incorporation, by-laws, stock transfer records and corporate seal

- Assignment of patents

- Certificate by the seller that all representations and warranties made by him in the Stock Purchase Agreement are true as of the closing date

- Receipt for a certified check or bank cashier's check for the amount of the sale

The escrow agreement, in this case, is the annual release of funds from an agreed-upon escrow amount to secure the buyer's interest that the warranties will prove to be as stated. The consultation agreement is an arrangement whereby the acquiring corporation retains the services of the former owner as a consultant, fixing the annual fee and the period of the agreement.

The purchaser in this transaction would deliver these items to the seller:

- Certified check or bank cashier's check for the amount of the sale, less escrow amount, payable to sellers

- Certified or bank cashier's check to a local bank, as escrow holder, if necessary, to cover the withheld amount

- Copies of resolutions of the Board of Directors authorizing the transaction

- Executed copies of the escrow and consultation agreements

- Opinion of buyer's counsel that the purchaser is a duly organized corporation in good standing; that the agreement has been duly approved and authorized by the purchaser; and that the agreements are valid, legal and binding obligations on the purchaser

- Letter from the purchaser that he is acquiring seller's stock for his own account and not for resale

- Certificate that representations and warranties made by the buying corporation in the Stock Purchase Agreement are true on the closing date

- Receipt for the seller's stock certificates and stock book

The bank, if represented, will deliver its receipt for the purchaser's certified checks made out to it for escrow purposes.

Action to be taken following the closing would include transmittal by the purchaser of legal documents transferring patents to appropriate government officials for recording.

Rigorous attention to detail is required in preparing for and executing the closing, to avoid last-minute hang-ups. As an example, in one closing the seller's legal counsel discovered, early in the proceedings, that the acquiring company's check was not a *certified* check; this had been overlooked by the acquiring company's own counsel. The president of the acquiring company had his treasurer (both were present at the closing) go to the bank and get a certified check. Fortunately the bank was still open for the day.

Because of the importance of the event and the need for minute attention to detail, tension usually mounts during the exchange of documents. When finally the buyer's check emerges from under the pile of documents and is presented to the seller, his comment is usually to the effect, "Wow! I was wondering when we'd get to this!"

And with that, "his" company is merged with the purchaser's.[3]

---

[3] A question that arises in a stock sale is the matter of who gets the profit made by the selling company between the date of the Purchase and Sell agreement and the time of closing. The answer is that if the seller warrants the balance sheet as of a certain date, then the buyer gets the operating profit; but this is a matter of negotiation between the two parties.

CHAPTER **13**

AFTER THE MERGER

What will happen to my company after the merger?" This question obviously comes up long before the merger. It is, in fact, a primary consideration in deciding *whether* to sell. A corollary question is, "What will *I* be doing after the merger?"

These post-merger considerations relate to "why sell," and close a loop that began with the owner's analysis of reasons for and against selling.

In most mergers the seller maintains some connection with the company, either in an active capacity or as a consultant. The basic change is going from owner-manager to manager. If he leaves the business, however, for retirement or other reasons, he can wash his hands of it—"Well, that's that"—and walk away from this part of his life. On the other hand, whether he stays or not he may have a very human feeling toward the employees, customers, suppliers and other associates who have been his main contacts. Because of them another question looms large in his thinking: What will happen to the people after he sells the company?

This chapter is not designed to analyze post-merger problems; rather, it portrays briefly what the seller can expect will happen to his company after he sells it, since this bears on his "sell" decision.

OPERATIONS

One of the first actions after a merger is a thorough look at the acquisition by the parent company, particularly in accounting aspects, to make sure the ship has not sprung leaks that were not previously detected. Prior to merger this kind of investigation is restricted because of time, security and the relationship between the two organizations. Afterward these restrictions are removed, and the purchaser will be intent on maintaining and enhancing the profitability of the newly acquired operations.

The purchaser will move quickly to firm up relations with key customers and suppliers, assuring them that the merger will not impair the relationship built up in the past.

He will probably institute economies, cutting back proposed expenditures for capital equipment, product development, EDP expansion and other costs that can be postponed without hurting immediate profits. On the other hand, he may give a "let's go" signal, because the seller naturally wants the newly merged company to grow.

The acquiring company's operating procedures and practices usually are more formalized and more detailed than those of the merging company. Recognizing the relative ease of administration when uniform procedures are in effect throughout the corporation, some purchasers move swiftly to incorporate their policies and practices. But the more enlightened purchasers recognize that they too have a learning process after the merger, and they take time to learn the reasons for the merged company's practices before considering changes.

One of the major operating benefits that can reasonably be expected of the merger is access to the finances needed for growth. For this reason alone the owners of many selling companies are glad they made the "sell" decision. Corporate finances can make the difference between unleashing the company's full potential and continuously struggling against more affluent competition in terms of market penetration, key people, facilities and new products development.

PLANNING AND CONTROLS

After the initial transition in operations, the company's objectives will get particular attention. Often the acquired company's objectives prior to merging are based on short-term opportunity or are overly restrictive. For probably the first time in its history, the merged company's objectives, capabilities and business opportunities will be integrated, like three-dimensional axes, to determine the areas where these fundamental factors coincide:

Figure 13-1. Fundamental Business Factors.

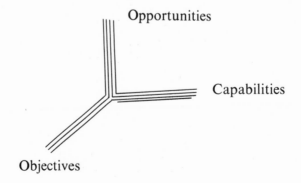

Opportunities

Capabilities

Objectives

In implementing these objectives, planning will probably be more long-range (that is, three to five years in the future rather than short-term) and will be the basis upon which management performance is evaluated.

The purchasing company will often install more effective cost controls, since many smaller companies frequently do not know the cost of individual products or services. One major acquisition-minded company bases its evaluation of a merger partner primarily on *gross*, rather than *net*, margin. Through experience this company knows that if the gross is up to the industry average or better, the company can effect enough overhead control measures to ensure a good net.

PEOPLE

How and when employees hear about the merger is extremely important. In professional handling of the transaction, there will be little or no publicity until the appropriate time. Then, at the earliest suitable moment, it is mandatory that employees be told of the event before getting the news from outside sources. Employees will be satisfied with the situation and will continue with the company if they are told early enough and by the top man.

If employees first hear the news by rumor or by public news media, however, they may well feel they are being herded from one corral to another and that they had better stampede out of the situation. Among those who may cut out are the very people the acquirer most wants to retain. These individuals may conclude that if something good were going to happen to them, they would have been told about it; not having been told, they had better look for greener pastures.

In telling employees about the merger, the gist of what the seller says may be, "I've sold the company. But I'm staying with it and there won't be any significant changes in operations." Or, "I wanted the company to grow bigger, and had your interests in mind. By myself I could only grow so much; but with the sale, you can grow to higher levels." This should be accompanied by an opportunity for a question-and-answer session between employees and the seller.

All employees will naturally wonder how the sale of the company will affect their income and employee benefits, and the seller should ascertain the purchaser's intentions on these items prior to the transaction. Usually these matters are stated during the merger announcement. Often the higher benefit package prevails, whether buyer's or seller's, although this is a matter of negotiation. In an earn-out there would probably be no change in benefits.

Acquisition-minded companies are increasingly aware that a going organization is the main factor in achieving a merger's potential, and they now approach the transition with far more skill than in the early days of the merger surge. In fact, the more observant acquirers will evaluate the seller partly on the extent of planning he has done for this transaction; this indicates to them something about the quality of management in the company and the extent of post-merger problems they may encounter.

The scars of merger, for both individuals and the company, can last a long time unless the seller and the purchaser give detailed thought to how the integration will be accomplished. A seller who is concerned about what might happen to key employees should find out ahead of time what the acquirer plans, and what he has done in previous purchases. Then he should take action to secure their positions or help place them, if he feels this degree of loyalty.

THE OWNER RETAINED

Finally, and of particular interest, is what the former owner can expect if he remains as a key employee in the merged organization. This depends primarily on knowing the purchaser and knowing, deep down, what his own intentions are: whether he is enough of a professional manager to lead his operation to new levels of growth, or whether he will "go slack."

The best indication as to how he will fare comes from talking to others who sold their companies to the same purchaser. The more experienced purchasers recognize that the head of the merging company is the key individual in bringing the operation up to the performance level expected; his influence is critical in maintaining a competitive, hard-hitting company team. If the top man gives wholehearted support to the new operation, this attitude will permeate the whole organization. (His support often depends on whether he is on an incentive basis or just on salary.)

Because his attitude is so important, the seller-manager may find that—at least for a time—he can retain his car, club memberships and other privileges of office.[1] Most sellers and acquirers recognize that the status symbols and privileges of office are particularly important to the top executive. The success of an acquisition can usually be measured by ability to motivate top management.

A particularly difficult aspect of remaining in the organization, however, is that the former owner and ultimate decisionmaker is now an employee; he must refer certain matters to his new superiors and, naturally, must carry out the policies and desires of corporate management. *He should ascertain prior to the merger exactly whom he will report to, and his own authority and responsibilities.*

Many former owners are unable to make this adjustment and leave within a relatively short period. But there also are those who welcome the opportunity to grow in stature and influence in a larger organization, and for them the merger is a breath of fresh air. One of the largest of the diversified companies moves these capable individuals up the ladder by using their experience and talent for, in turn, searching out and acquiring new merger candidates.

---

[1] "When one company is absorbed by another, I do not think that the importance of maintaining individual status and prestige can be over-played. In my experience it is easier to knock $20,000 off the purchase price of a company than it is to get a man to give up his car." Willard F. Rockwell, Jr., "How to acquire a company," *Harvard Business Review*, September-October 1968, pp. 121–132.

CONCLUSION

Sometimes the seller has second thoughts as to whether selling was the best decision, and whether he might have negotiated a better deal. The decisionmaking process in a merger is the same as in any other important aspect of business. If the owner has gone about the merger in a professional way, from "why sell" to operations after the merger, he can be assured he made the right decisions and took the right action.

The rest is easy: Enjoying some of the best years of his life, in or out of the company.

LEONARD McBIRNIE

Leonard McBirnie was the principal owner of McBirnie Castings, a small firm ($1.5 million in sales) that he founded in 1955. Because of business and personal reasons, he decided to sell the company if he could get the price he wanted.

The purpose of this appendix, an actual case, is to illustrate the steps leading to the sale of a company, and the effect of terms on the value received by a seller.[1] It also illustrates, in abbreviated form, the major documents involved in consummating a merger.

BACKGROUND

Leonard McBirnie had been the general manager of a castings company. He was in his late forties when he decided in 1955 to resign his position, pool his resources and start his own firm, McBirnie Castings Company. Shortly afterward he allowed two former associates, Marvin Wilson and Jerry Hruska, to join him, giving them the opportunity of buying part interest in the company. Over the next decade the business grew to well over $1 million in sales; it was profitable enough for the two associates each to acquire 10 per cent equity, paid out of bonuses earned.

In the late 1960's, however, McBirnie's brother died and left both a substantial inheritance to McBirnie and the time-taking job of unraveling a complicated estate. McBirnie was then almost sixty, and the combination of these personal factors and a decline in the industry in 1967 and 1968 caused him to consider selling his business, after discussing the situation with his two partners.

McBirnie had no knowledge about how to sell the company, however. Recognizing the need for assistance, he asked his accountant, who was also a close personal friend, to recommend a specialist in mergers. The accountant recommended a merger packager, who was contacted and who then met with McBirnie and the accountant. Subsequently McBirnie arranged for the merger packager to represent the company in obtaining a logical buyer.

---

[1] Company and individual names are disguised in this case and in the following case histories in Appendix A.

LOCATING A BUYER

Locating a logical buyer had one illogical hang-up: McBirnie wanted a million dollars for the company although it lost $19,000 the previous year, and its total net worth was $240,000. The merger packager learned that McBirnie had developed some new processes during the year. These had required substantial expenditures but promised to yield a good profit in the future. Accordingly, he suggested that McBirnie be willing to accept an earn-out on the basis of proving this earnings potential to a company that would be willing to pay the asking price. McBirnie and his two partners agreed to consider an earn-out plan.

Soon afterward the merger broker introduced a prospective buyer, Dave Edwards, who was president of a subsidiary of a diversified company. Edwards and McBirnie had similar business backgrounds and a number of mutual acquaintances, and the "chemistry" seemed right for a merger. Edwards was impressed by the cleanliness of the plant and the company's inventory; and he then had his financial partner review accounting and fiscal data in preparation for making an offer.

After further analysis and investigation, Edwards made the following offer: (1) Stock in his company equivalent in value to McBirnie Castings' net worth, with a two-year guarantee in case the stock declined in value during that period. (2) A provision that the sellers could earn an additional $250,000 in stock over a five-year period, at the rate of $50,000 per year, as long as the company showed earnings of $100,000 per year before taxes.

Edwards gave the following rationale for his offer. McBirnie Castings' business was tied to the fluctuations inherent in its industry, and, moreover, was largely dependent on a single customer. In addition, it was considered to be a job shop, although a very good one. Also, the firm had created an unusually liberal profit-sharing plan for its employees.

After discussing the proposal with his two partners, McBirnie rejected the offer outright.

During the next two months the merger broker presented the company, on a nondisclosure basis, to several potential acquirers. None proved to be sufficiently interested. Then he learned that Astral Industries, a diversified corporation, had acquired a company that produced castings of a type that complemented the products of McBirnie Castings Company. He contacted the chairman of Astral, who expressed an interest in McBirnie. The broker advised the chairman that the owners would not consider a price of less than a million dollars, including an earn-out. The chairman was still interested. At a meeting between McBirnie and representatives of Astral's subsidiary, it developed they were old friends and had a high regard for each other.

Within a week the chairman of Astral authorized the merger packager to make the following offer to McBirnie. First, Astral would pay $475,000 for the assets of McBirnie Castings. This would be payable 29 per cent on closing with the balance to be paid over a five-year period, with 7 per cent interest on the unpaid balance. Payments would be made annually, on the anniversary of the closing. Second, Astral would assume and pay all debts, liabilities and obligations shown on the McBirnie Castings Company balance sheet. And third, Astral would issue its common stock to the sellers in an amount equal to $2 in stock for every dollar of after-tax net earnings of McBirnie Castings, provided the company earned $40,000 or more per year after taxes. This earn-out was good for three years, with the potential of earning up to 40,000 total shares of Astral stock. In computing after-tax earnings, Astral would deduct from McBirnie Castings' earnings the applicable Federal, state and franchise taxes, and a one per cent management charge. The market value of Astral stock (traded over the counter) would be figured in this way: Its price would be computed as the average of bid and asked prices for each of ten trading days preceding the annual date of delivering the stock to the sellers; but the market value, for purposes of earn-out, would not be greater than $20 per share on the first delivery, $30 per share on the second delivery, and $40 per share on the third anniversary delivery date. In no event, however, would market value be less than $10 per share. The effect of this earn-out provision would provide great incentive (to Hruska and Wilson, who were staying with the company), and protection for both buyer and seller against excessive swings in the value of Astral's stock.

The sellers analyzed the offer in detail, with the merger broker explaining the effect of taxes, form of transaction and earn-out provisions. They analyzed the outlook for Astral, whose stock was then selling at $12. Hruska and Wilson, who were both in a 50 per cent income tax bracket, saw that by eliminating profit-sharing in favor of greater earnings, they would receive *two dollars* in stock for every dollar of net profit, rather than *half a dollar* (after personal income tax) on each profit-sharing dollar earned. Because of this multiple of four, they elected to eliminate the profit-sharing plan.

McBirnie and his two partners then accepted the offer, and attorneys for the two parties were instructed to draw up final papers. All agreements were signed, and approval of the state Corporation Commissioner was requested. As a footnote to the intensity of merger activity in general, final approval for the merger was not obtained from the Commissioner until almost four months after it was requested. Closing took place a few days after approval.

INVESTMENT

It is interesting to note the value of Astral stock as an investment of Leonard McBirnie. Within three months after the transaction, Astral was listed on the American Stock Exchange, and its stock went up from $12 to $25 per share. If it remained at this level and if McBirnie Castings earned the minimum requirement of $40,000 net after taxes, Leonard McBirnie would receive 4,000 shares of Astral, based on the $20 per share market value provided in the agreement for the first anniversary (that is, $2 in stock for each dollar of the $40,000 net, or $80,000 in stock, divided by $20 per share.) Meanwhile, the stock would have risen to $25 (or probably higher), giving McBirnie an additional $20,000 paper profit. The stock could, of course, have gone down; but in this case he would have received additional shares to offset the declining price.

MAJOR DOCUMENTS IN THE TRANSACTION

The major documents and their provisions in the McBirnie merger are indicated below.

Promissory Note. The promissory note for the balance of 71 per cent of the $475,000 paid for McBirnie Castings' assets was payable at 7 per cent interest on the unpaid balance, with both principal and interest due and payable in eight semi-annual installments, beginning June 30, 1969. The note provides for prepayment of principal or interest at any time without penalty.

Acquisition Agreement. After the "whereas-es" describing the intent of the transaction and citing approvals of the respective boards of directors, the agreement defines the terms of the merger. *Section 1* provides for the transfer of assets and property from McBirnie Castings to Astral, and assumption and payment by Astral of outstanding liabilities. *Section 2* provides for consideration: delivery of cash, notes and Astral stock as described in Astral's offer. *Section 3* pertains to the closing, and actions to be taken by each party at the closing. *Section 4* provides for representations and warranties of McBirnie Castings: corporate status, capitalization, financial statements, agreement not to make significant changes between the date of the agreement and the date of closing and warranties as to accounts receivable, inventory, taxes, litigation and correctness of warranties on the closing date; and agreement that McBirnie and his two partners will hold Astral's notes to be received as an investment rather than for resale. *Section 5* sets forth Astral's representations and warranties, including statements as to its organization and approval for executing the agreement, verification of financial statement and delivery of voting stock to the sellers. *Section 6* consists of a long series of covenants of McBirnie Castings: actions prohibited pending closing, conduct of business pending closing and other agreements on the part of the sellers. *Section 7* lists Astral's covenants: agreement to hold McBirnie data and information in confidence and to sign the Registration Agreement (pertaining to registration of Astral stock) at the closing. *Section 8* cites "conditions precedent" for the benefit of the sellers; namely, that Astral's representations and warranties will continue in effect to closing, and that it will take all actions necessary to effect Astral's obligations under the terms of the Acquisition Agreement. *Section 9* cites similar conditions for the benefit of Astral. These include performance of covenants and warranties by McBirnie, delivery of certificates as to fulfillment of conditions, absence of litigation, material changes since the agreement (including obsolescence of products or service, labor dispute, resignation of key personnel, accident, flood, drought, civil disturbance or act of God in "material changes") and receipt by Astral of McBirnie counsel's opinion. *Section 10* provides conditions precedent for the benefit of both parties, such as securing a permit from the state Commissioner of Corporations. *Section 11* provides that Astral will not be subject to a finder's or broker's fee. *Section 12* provides, "All statements contained in any exhibit or other instrument delivered by or on behalf of the parties hereto, or in connection with the transactions contemplated hereby, shall be deemed to be representations and warranties hereunder. All representations, warranties and agreements made by the parties to this agreement or pursuant hereto shall survive the closing and any investigations made by or on behalf of the parties." Related to this is *Section 13, Indemnification,* in which each party agrees to indemnify the other for any breach of warranty, guarantee or other commitment. The remaining sections pertain to notices in writing, the agreement and amendments, successors and assigns and agreement that the law of California (McBirnie Castings' state) shall govern.

Letter of Agreement from McBirnie Castings Stockholders. This is a letter in which each of the selling stockholders agrees that, on receipt of Astral stock, he intends to hold the stock, except as permitted by the Securities Act of 1933.

Release and Assignment. This is a transfer, assignment, and so forth of each McBirnie stockholder's rights to inventions, patents, copyrights, formulas, manufacturing methods, and so on to Astral.

Agreement Re Employment and Competition. This clause, requested by the sellers, provides for an employment contract for Leonard McBirnie's partners, Marvin Wilson and Jerry Hruska, to continue with Astral's subsidiary, citing the compensation, terms of employment, services to be rendered, agreement not to associate with a competitor, and so forth.

Registration Agreement. The registration agreement pertains to registration statements that may be filed by Astral regarding its securities, and the rights and obligations of shareholders (such as McBirnie) in regard to these actions.

ARTURAS PLASTICS COMPANY
A Divestiture

As indicated in Chapter 4, the divestiture market is becoming increasingly active. The Arturas Plastics Company is a sell-off case in which the company had been bought by a listed corporation (Mercury Chemical Company) and subsequently was put on the market, the stated reason being that Arturas no longer fitted in with the parent company's plans. The case is noteworthy because of the sequence of events leading to the sale and the interpersonal relations among the personalities involved. Actual names of individuals and organizations are, of course, disguised.

Cast of Characters

Mercury Chemical Company—Seller
  Howard Daniels, President
  Bill Levy, Vice President, Finance
  Robert Miranda, Controller

Arturas Plastics Company—subsidiary being spun off by Mercury

  Robert Josephson, President

Raymer Chemical Corporation—a Midwestern prospective purchaser of Arturas

  John Ray, President

Sidney Trevor—a prospective buyer for a diversified corporation

Duane McCoy and Carl Boehm—independent purchasers

Peter Marchand—a merger broker

BACKGROUND

Early in 1967 Peter Marchand received a phone call from Bill Levy, VP Finance of the Mercury Chemical Company of Seattle. Levy explained that Mercury had decided to sell its plastics division, Arturas Plastics Company located in Los Angeles, and wanted a competent merger broker to handle the sale.

Mercury had acquired Arturas Plastics in 1958 because at that time Mercury wanted a source of containers for Mercury's garden fertilizers; it had appeared that a new process called blow molding was going to replace glass, and Mercury wanted a protected source. Now, in 1967, it was recognized that blow molding was not effective for Mercury's particular container requirements and that, although Arturas was profitable, it took too much of Mercury's management time. Consequently the corporation had decided to spin it off. Arturas' volume was about $7.5 million in

Figure A-1. Profile

Products and Services
This company has three basic departments: The Injection Molding Department produces both proprietary and custom parts; the Compression and Transfer Molding Department primarily produces a line of proprietary products with some custom parts added; and the Blow Molded Department produces containers to customer requirements. Approximately 40 per cent of the business is proprietary and 25 per cent is commercial. Over $3 million annual sales are in plastic parts for aircraft and missiles.

Management and Personnel
The present operating group, in general, has been with the company for five years or more and will stay with the operation. The personnel includes a complete staff for the tool department, which is a very valuable asset in this day of tight labor. The company employs approximately 450 people.

Ownership
For the past nine years this company has been owned by a large, publicly-held company headquartered outside of Southern California. The management of the parent organization has determined that even though this is a profitable division, it does not fit the total corporate program, and therefore it is in the best interest of the company to dispose of this operation.

Plant and Equipment
The plant is located in one of the better industrial areas of Los Angeles and is a single story, 204,000-sq.-ft. building with a good lease arrangement. Some of the latest injection molding equipment was installed in the first quarter of 1967. The plant is supplied by both steam and refrigeration for the presses. The capacity is approximately double its present production.

Terms and Conditions
The owning company will sell inventory with fixed assets of this business for $4.5 million.

Figure A-2. Operating Data

	6 Months 1968	1967	1966	1965	1964
Net Sales	$3,855	$8,121	$6,105	$5,763	$6,759
Costs of Goods Sold	3,012	5,688	4,617	4,641	6,528
Gross Profit on Sales	843	2,433	1,488	1,122	231
Deductions from Gross Profit	54	96	42	159	321
Gross Profit After Deductions	789	2,337	1,530	963	(90)
Operating Expenses	567	1,203	1,047	1,194	1,065
Operating Profit	232	1,134	483	(231)	(1,155)
Nonoperating Income and Expense	12	60	156	90	597
Profit Before Taxes	234	1,074	327	(321)	(1,752)
Provision for Federal Taxes	81	483	120	(201)	(1,035)
Net Profit	153	591	207	(120)	(717)

Figure A-2. Pro Forma Balance Sheet

		In Thousands
Current Assets		
Cash	$ 102	
Receivables	819	
Inventories	2,175	
Prepaids	111	
Total Current Assets		$3,207
Fixed Assets		2,322*
Total Assets		$5,529
Liabilities		
Payable	$ 300	
Accruals	207	
Taxes	63	
Total Liabilities		$ 570
Worth		4,959
Total Liabilities and Worth		$5,529

*Net of $2,352,000 accumulated depreciation.

	1967	1966
Gross Sales	$8,229	$6,231
Less Returns	108	126
Net Sales	$8,121	$6,105
After-Tax Profits	$ 591	$ 207

1966 and estimated at $8.1 million for 1967, with an after-tax profit of $600,000. Levy stated that Arturas had a capable management organization, that Mercury would not consider a sale for less than Arturas' book value because of pending arrangements with financial sources and that only a cash deal would be considered.

With additional information subsequently obtained, Marchand developed the profile and fiscal information shown in Figures A–1 and A–2.

LOCATING A BUYER

Marchand knew that a Midwestern company, Raymer Chemical Corporation, was looking for a plant in the Los Angeles area. After telephone discussions with John Ray, President of Raymer, a meeting was arranged in Los Angeles between Ray, Levy, Marchand and Robert Josephson, the President of Arturas. As Ray was being shown through the company, it became apparent to Marchand that other prospective buyers had been in the plant previously, and that although Josephson had not before been apprised of the reason for the meeting, he was not at all surprised when informed of the purpose.

Ray made as detailed a look at sales, facilities, product line, financial reports, and so on as he had time for. Marchand provided him with an equipment appraisal and an audited three-year financial statement, and Ray said: "It looks okay to me. I want to have my VP for Manufacturing come out and take a look at it."

The following week Ray's Vice President for Manufacturing arrived. Marchand introduced him to Josephson, and left him in Josephson's hands for a detailed, two-day inspection of the plant.

In the meantime Marchand had learned from Josephson that there had been serious negotiations with a local diversified firm for almost a year, but that for some reason negotiations had broken off. Levy later explained that the prospective acquirer felt the timing was not right to get into the plastics business.

A week later Ray called Marchand and advised him that Raymer was not interested in Arturas. He stated there were two reasons. First, the equipment was antiquated, and in order to meet Raymer's exacting specifications for an operating division, all the equipment would have to be brought up to date. Second, Raymer felt that there was not sufficient depth of management to justify the expenditure for a company of Arturas' size and distance from Raymer headquarters. Marchand thanked Ray for the response and said he would pass the information to Levy, which he did.

In the next several months Marchand presented the corporate profile to other prospective buyers, but for various reasons none were interested. He also kept in touch with Josephson; on one occasion Josephson said, "Arturas could never succeed as long as we're with a chemical company, because they don't understand our operation at all . . . . "

Marchand then had a meeting with Sidney Trevor, who was interested in acquiring a company in the chemical or plastics field for the corporation he represented. Marchand arranged for Trevor and his financial backer to meet with Marchand and Josephson. After an all-day meeting Trevor asked Marchand for detailed information on Josephson. Marchand gave him the following information: Josephson was a graduate engineer and had spent his whole life in the plastic business. He had taken over direction of Arturas two years before, and reduced a loss to $270,000 the first year, and showed a $260,000 profit the next year. In the current year he was confident of a $590,000 profit. Josephson had mentioned that he had some problems with Mr. Levy, in that Levy had not lived up to some of his commitments; but he, Josephson, was confident that he could negotiate a satisfactory employment contract with Arturas' buyer.

A few days later Josephson advised Marchand that a large Eastern firm had been interested in acquiring Arturas but had later called off the deal. Josephson also mentioned that Bill Levy was resigning from Mercury, after thirty-five years, to become president of another company.

Marchand immediately called Levy, who confirmed that he was leaving and would be succeeded by Bob Miranda, Mercury's comptroller. Levy also said that Mercury was going to fire Josephson "because of some things that are going on." He said that because he had hired Josephson, it was his duty as well as his decision to fire him; and that he would do this within the next few days.

A few days later Levy fired Josephson. It developed that the main reason interested companies had been shying away from Arturas was Josephson. He had been telling Raymer, Trevor and others before them how important he was to the company's success, and that his terms for staying with Arturas were a high salary, percent of pre-tax profit and other conditions that were unreasonable in total.

After Josephson left the company Trevor spent several days with Arturas' Vice President, who had been with the company for fifteen years and was provided with statements of inventories, appraisals and other information. Trevor then asked for a meeting with Peter Marchand. In this meeting he told Marchand that the company was in far worse condition than he had thought. Profits, he said, were achieved by inventory accounting techniques; the inventory was overstated and largely obsolete, and the equipment was in poor condition and would have to be largely replaced. He stated that he was not interested in buying the company, but that he was willing to explain to the Mercury people how serious the situation was.

A few weeks later, when Daniels and other Mercury executives happened to be in Los Angeles, Marchand arranged for Trevor to present his analysis to them. The meeting was interesting for shock value, if nothing else. Trevor was almost brutal in his appraisal. He concluded his talk with the Mercury people by saying that he would not take Arturas as a gift, and he thought their only salvation was to liquidate it and take their lumps. He concluded with, "Gentlemen, you have * * * *! this company up, and it's your baby!" Exit Mr. Stanley Trevor.

NEGOTIATING THE DEAL

Two weeks later Marchand had a meeting with Duane McCoy and Carl Boehm, who had a small, successful service business and wanted to get into manufacturing. They could get financial backing and were interested in obtaining a problem company. After looking at the company and analyzing data, they concluded that they would be interested if proper terms could be worked out.

McCoy and Boehm arranged for a bank loan and a Small Business Administration (SBA) loan; they also raised money from friends. On the basis of this they made the following offer through Marchand: They would buy the inventory, machinery and equipment for $900,000 cash at closing; a note for $2.7 million bearing 7 per cent interest due in semiannual payments, with the balance due and payable in five years; and an additional note for $750,000 bearing no interest, with no payment to be made until four years later. The reasoning behind the second note was that the finished goods inventory included at least $750,000 in merchandise for which there was questionable value, because it was slow-moving. For one part in inventory there was a five-year supply, based on the previous year's sales; for another part there was an eight-year supply.

Negotiations proceeded, with Marchand meeting alternately with Howard Daniels in one room and with McCoy and Boehm in another room. Finally a definitive business agreement was reached as to price, terms, restrictions and operating arrangements. The buyers, their attorney and all parties to the transaction subsequently met in Seattle—twice—to finalize the Purchase and Sell agreement. Closing was set for thirty days later.

The closing took place in Los Angeles, two years after Marchand received the phone call from Levy stating that Mercury had decided to spin off Arturas.

How did McCoy and Boehm fare with their problem child? Extremely well. Josephson had bought excessively large quantities of raw material, at one-cent per pound savings; but this savings had been more than offset by labor and rework costs. The new owners therefore sold most of the inventory, at a profit, during the first six months and effected other stringent control measures. Because of these actions, their strong management backgrounds and a fortunate increase in the value of the company's holdings, they became, on paper, near-millionaires.

GAINES & BERGER, INC.

George Gaines and Silas Berger started a partnership in the 1880's as distributors of industrial supplies. The business prospered for almost seventy-five years, until two third-generation brothers, Joseph and Benjamin Gaines, became President and Vice President of the Company. During the sixteen years of their management, sales continued to grow slowly but profits were erratic. Profit on sales was almost zero in 1967 and 1968. By this time the two brothers were sixty-nine and sixty-five years old, and because of the profit picture and other reasons they decided to sell the assets of the company.

The Gaines case is noteworthy because it is typical of problems sometimes encountered in the sale of a family-controlled business: the decision to sell, loose management in normal operations and in the merger transaction and retaining the company's regular attorney to represent it in the merger. In this case a sale of assets at 50 per cent of book value was a good deal for the sellers.

BACKGROUND

About the turn of the century, Silas Berger, one of the two original partners, died, leaving no heirs. George Gaines acquired his partner's interests and continued as head of the growing business until 1912, when he turned it over to his only son. For the next forty years the son was very active as chief executive, until he in turn was succeeded by his son, Joseph Gaines.

Joe Gaines had worked up the company ladder through shipping, warehousing and buying before becoming President; and his younger brother, Ben, had also grown up in the company. A third brother was inactive in the business, but was also a major shareholder. The remaining outstanding shares were held by thirteen children of the three brothers; some of these fourth-generation members were also employed in the business.

THE DECISION TO SELL

As early as 1963, Joe and Ben Gaines had talked about selling the business, and they discussed this possibility off and on for five years. Because of the family connection with the company and the fact that they made a comfortable living, they were reluctant to come to a "sell" decision.

By 1968, however, several factors influenced them to sell the business. One was that their oldest competitor in the business of distributing industrial supplies had gone rapidly downhill in the past several years by failing to apply computerized inventory control; then, too late, the competitor had over-reacted to the extent that he had been forced to close his doors and liquidate. This weighed heavily on the Gaines' thinking. In addition, employees in the plant had been agitating for union recognition, and were successful in getting the Teamsters and Warehousemen's Union as bargaining agent; this bothered Joe Gaines considerably, because he had known many of these people since he was a child and now they were looking elsewhere for support. Another factor was that although Ben Gaines had two sons who were active in the business, none of Joe Gaines' four sons-in-law were in the company.

Another reason for selling was the company's results in the past five years. Sales had grown modestly, but the profit picture was very unsteady. For 1967 and 1968, net after taxes had been less than two-tenths of one per cent on sales, and only one per cent on assets.

Finally, the company had a bank loan for operating purposes which it was supposed to clear annually. But by 1968 the loan was at the $1 million maximum and had not been cleared for three years.

FINDING A BUYER

The Gaines brothers had no experience in merging or selling a business, and asked their labor consultant to recommend a merger specialist. The specialist then met with the two brothers, and after this meeting a letter agreement was signed whereby he agreed to serve as merger broker for them.

In preparing a profile on the company, it was apparent to the specialist that the company's real estate had appreciated greatly in the past several years, partly because of general location and partly because a freeway had been built within half a mile of the company property. The property also had a rail siding and a large truck dock.

The Gaines brothers set an asking price of $3.7 million if the sale included this real estate, or book value if real estate were not included.

The merger specialist discussed a list of likely prospects with the two brothers. Then he approached the most likely buyers by letter, with a corporate profile attached. In complete unison, every one of these prospects turned down the prospective merger on the grounds the company was in a dying field. Additional companies were contacted; they also turned it down.

The owners reappraised the asking price and modified it to $1.5 million dollars, exclusive of real estate.

About this time a firm representing a major liquidator made an offer for the company, as a going business. In hopes that they could make the company profitable, the management offered $2.6 million for all the company's assets, provided the Gaines' retained the corporate entity and paid off all liabilities. Since liabilities totalled $1,850,000,000, this meant the owners would receive only $750,000 for their company. After careful analysis the brothers replied that they would sell the company's assets for $3,450,000, or $850,000 more than the offer. The prospective buyer would not consider going this high, however, and the two parties did not pursue further discussions.

At this time the merger specialist met William Schwartzstein, Chairman of Detroit Industries. Schwartzstein was about to obtain control of a publicly-held company and was seeking to build it into a larger company through mergers and internal growth. Both Schwartzstein and the company President, David Morton, were capable, hard-driving executives, and both had experience in building sick companies into profitable, listed corporations.

Several meetings were held with the principals of Detroit Industries, on a non-disclosure basis (that is, without disclosing the name of Gaines and Berger, Inc.); and the merger specialist then obtained Joe Gaines' approval to present the company to the Detroit executives. A plan was worked out for acquiring Gaines and Berger's assets, with 29 per cent initial payment and an installment payout. The specialist then brought the plan to a meeting with the Gaines brothers. He gave them the background of Detroit Industries and its key individuals, and then explained the benefits and possible risks of the Detroit proposal. Joe Gaines said he would like to study the figures at home, but could not get at it that evening because of a bowling engagement or the next day because of a golf game; but he would try to have a response ready the following day.

NEGOTIATIONS

When the Gaines' had analyzed the proposal, their response was that if they planned to live without spending capital, they would need at least another $250,000 in addition to rental from the property, which they planned to retain. The merger specialist replied that the buyer might come up with another $100,000, but that this was probably the extent of his negotiating range.

Mr. Schwartzstein, traveling to Cleveland, Bridgeport, Conn. and Washington, D.C., on another merger, was contacted. The specialist suggested that in addition to cash, 15,000 shares of stock in Detroit Industries be included on a "put and call" basis to add another $100,000.

In a long meeting with the Gaines brothers, the specialist discussed Detroit stock and explained puts and calls. Then, because of their apparent hesitation about accepting stock, he suggested that they could pay his brokerage commission with the stock so that they would not be affected by fluctuations in its value. This seemed acceptable to the owners, who would receive $1.4 million net, with 29 per cent payable at closing. They would also receive approximately $100,000 annual rental on the real property.

Joe Gaines spent several sleepless nights, however, worrying about the security of the remaining 71 per cent due in subsequent payments. What would happen if

Detroit Industries decided to go out of business and liquidated their assets? How would the shareholders of his company get the remainder due?

The merger specialist assured Mr. Gaines that he could secure payment by an interest in receivables, payable to the bank, and also that the Gaines' attorney would make sure they were protected. At this point the Gaines brothers agreed that if their attorney concurred, they would go through with the deal.

The Gaines' lawyer for the merger was an elderly, personable gentleman who was their regular company lawyer. In a meeting of the lawyer, the Gaines brothers and the merger specialist, it developed that the two brothers would be satisfied only with cash, payable in full within one year. At the conclusion of the meeting the specialist impressed on the others present that time was of the essence, since Detroit's acquiring the company depended on closing the deal in time for Detroit to move another operation into the Gaines-Berger facilities. It was in this way that Detroit expected the merger to provide an operational profit. The specialist expressed the advisability of getting the attorneys for the seller and the buyer together without delay, to develop the Purchase and Sell agreement.

At a meeting a short while later between the parties, Mr. Schwartzstein said he would give a pledge of the unpaid receivables for the balance, after 29 per cent initial cash payment. Since the unpaid balance was $990,000 and receivables were well in excess of this (and would be paid weekly, up to 75 per cent of receivables, to the Gaines as received), and since the buyer would pay 7 per cent interest on the unpaid balance, the terms were apparently accepted by the sellers. At the conclusion, Schwartzstein asked, "Now is it all clear? Are there any questions? Do we have a deal?" There were no questions. It was agreed that the buyer's attorney would draw up a draft Purchase and Sell agreement.

By nine o'clock the next morning the buyer's lawyer had completed the draft agreement. The merger specialist took copies to the Gaines' office and to their attorney.

He was surprised a few hours later to get a phone call from the sellers' attorney saying that the sellers did not understand the deal, and that the specialist should talk with Joe Gaines. Gaines was not available until several days later; but at a meeting at that time, Gaines said their calculations showed they should have another $30,000 in order to have the income needed for retirement. The specialist advised him that he was sure Detroit would not raise their offer, and recommended that the Gaines brothers accept it.

The Gaines brothers discussed the merger further with their attorney, again going over points involved in a sale of assets, such as retention of capital stock and the corporate structure, and confirming that any tax benefits would accrue to them. The owners then agreed to sign, and the buyer's attorney drew up the final agreement.

CLOSING

It became increasingly difficult, however, for the buyer's attorney or the merger specialist to contact the sellers' lawyer, who was unavailable and did not respond to attempts to contact him by phone. Days lengthened into weeks. Closing did not take place at the appointed time, and the buyer became "cool" to the idea of ever finalizing the merger. Closing was delayed, in fact, for more than two months after the designated date.

Finally, in spite of the buyer's ebbing interest, the closing did occur. The sellers received cash according to the agreement, and more in rental of their property than they had made in profit during any of the previous five years. However, they were charged more than $10,000 for legal fees, considerably more than attorney's fees incurred by the buyer.

MADLOCK, INC.

The Madlock case illustrates the importance of inventory valuation, and how smooth a merger can be when knowledgeable people are involved. It also illustrates the extent to which the government may concern itself regarding restraint of trade. Madlock, a subsidiary of a large corporation, was doing about $8 million in sales at the time the Justice Department became interested in it.

BACKGROUND

K-Line, Inc., of New York had acquired Madlock from its founder in 1953 and continued to manufacture a line of industrial confectionery molds at Madlock's West Coast location. K-Line also later established a plant in Canada that produced a similar line of products.

By early 1969 K-Line was doing a total volume of $200 million annually, mainly in consumer products. The founder, Mr. Bertram Kay, was in the process of merging his company with a major international corporation. Combined sales of the two corporations would be in excess of $1 billion.

Shortly before this merger, however, the Department of Justice charged K-Line with restraint of trade in production of industrial confectionery equipment. When the parent company's merger was completed, corporate attorneys investigated the Justice Department's charges and instructed K-Line management to divest themselves immediately of Madlock's confectionery equipment line. K-Line did this, reducing Madlock's sales by $3 million. This action left Madlock operating below a profitable level. For this reason and the fact that all the other products of the corporation were commercial, whereas Madlock's products were industrial, the parent corporation decided to sell Madlock.

NEGOTIATIONS

Meanwhile the Burbank, Calif., plant of a major Eastern corporation, Sanders & Johnson, Inc., had received a large order to completely equip a Vietnam complex with movable partitions, metal work benches, and so forth. It desperately needed additional facilities similar to Madlock's. A merger broker arranged for a team from Sanders & Johnson's Pittsburgh headquarters to inspect Madlock and verify Madlock's capability to produce on a military contract.

Madlock's facilities were excellent, and a number of skilled operators had remained in the company even after the confectionery line divestiture. Consequently, Sanders & Johnson made the following offer to K-Line.

1. Sanders & Johnson would collect the receivables for K-Line and turn them over to K-Line.

2. Sanders & Johnson offered K-Line 75 per cent of the standard costs of inventory, including work in process.

3. They offered $600,000 for the machinery and equipment.

4. Sanders & Johnson also offered $2 million for the real estate.

K-Line accepted the offer on receivables, machinery, equipment and real estate, but made a counteroffer on the inventory: Sanders & Johnson would pay standard costs for all raw material and 90 per cent for work-in-process and finished-goods inventories. Sanders & Johnson's representatives responded by saying they would pay standard costs for the raw material, but would pay only 75 per cent for work-in-process and finished goods.

At this point, negotiators of the two companies sat down to work out the difference between 75 per cent and 90 per cent of work-in-process and finished-goods inventories. After a very keen bargaining session, the buyer and seller reached an agreement: 83 per cent of the inventories in question.

The agreement provided that Sanders & Johnson would pay $1 million at closing and the balance, $2.4 million, over a three-year period, with 6 per cent simple interest.

SMOOTHNESS OF THE TRANSACTION

K-Line and Sanders & Johnson both had in-house counsel who were experienced in mergers and acquisitions and who were responsive to the business purposes of these transactions. Because of this, the definitive agreement was drawn and a closing date was established without delays or difficulties.

The first contact between K-Line and Sanders & Johnson was in mid-July. As a result of the experience of the people involved, preliminary agreement was reached a month later, and the deal was closed in mid-September. This two-month period from initial contact to closing included the time required for a shutdown in order to made a complete inventory of Madlock property, equipment and materials.

APPENDIX B
REGULATORY ASPECTS

This appendix provides the owner of a closely-held company with a brief introduction to regulatory acts and agencies, since these may affect him directly; or, more likely, they may affect him indirectly by having an impact on his actual or potential acquirer. An experienced merger attorney can advise him in detail on pertinent regulations and regulatory agencies.

Regulatory aspects of mergers are likely to be of more concern to large, diversified acquirers than to small, closely-held sellers. The Federal Trade Commission, for example, keeps an eagle eye on prospective mergers involving companies having $250 million or more in assets, or who would have this magnitude of assets as a result of a merger. Regulatory agencies are not likely to give more than passing notice to the merger of a firm having $10 million or less in assets. Nevertheless, they may take action where the nature of the industry or other factors warrant closer scrutiny. In the acquisition of Fisher Brewing Company by Lucky Lager, the government initiated action even though Fisher's assets were only about $2 million.

The Federal agencies most likely to be concerned with merger transactions are the Federal Trade Commission, the Justice Department's Antitrust Division, the Securities and Exchange Commission and the Internal Revenue Service. Other Federal agencies may have an interest where a merger involves such commercial businesses as banks, trust companies, communications and airlines. In addition, state laws and regulatory agencies are a consideration in merger transactions.

FEDERAL TRADE COMMISSION
AND THE DEPARTMENT OF JUSTICE

In 1890 Congress, concerned that industrial giants were sapping the strength of competitive American business, passed the Sherman Antitrust Act. This act declared that every contract, combination or conspiracy in restraint of trade or commerce is illegal. The Justice Department was charged with enforcing the act, and did so with gusto.

Later, however, Congress recognized there was a serious weakness in the law; namely, great damage could be done during the period, perhaps several years, required in smashing a monopoly. For this reason Congress passed two new laws in 1914; one was the Clayton Act, designed to stop business practices, including mergers, that *threatened* to restrain trade; the other established the Federal Trade Commission (FTC), whose purpose includes seeking out and destroying monopolistic tendencies *before* they can become a market menace. The FTC is generally involved in test cases from which merger guidelines will develop.

The FTC and Antitrust Division of the Department of Justice have concurrent jurisdiction over initiating action on merger cases, and have a liaison arrangement for clearing with each other to avoid duplication. This results in some specialization, particularly as to certain industries in which each has built up substantial experience and expertise.

Generally these agencies hear of acquisitions and mergers through regular news media, such as financial pages and business journals, or through complaints by minority stockholders in companies whose stock is being acquired. Both are concerned with conglomerate and vertical mergers, as well as horizontal ones. Each will give an advance ruling on a proposed merger, if requested to do so; this usually takes three months or less.

If a complaint results from the agency's study, a complaint by the Justice Department is tried in a Federal District Court; if either party appeals, the case goes to the U.S. Supreme Court. A complaint by the FTC goes to a FTC examiner; an appeal from his judgment goes to the Commission, who may agree or disagree with his decision. Appeals beyond this go through courts of appeal, and can eventually go to the U.S. Supreme Court.

INTERNAL REVENUE SERVICE

The Internal Revenue Service (IRS) is concerned with tax aspects of the form of merger, discussed in Chapter 10. Strict compliance with the IRS provisions is required in each form of reorganization.

Section 368 of the 1954 IRS Code defines corporate reorganizations within the tax-free provisions as being:

(A) A statutory merger or consolidation;

(B) The acquisition by one corporation, in exchange solely for all or a part of its voting stock (or in exchange solely for all or a part of the voting stock of a corporation that is in control of the acquiring corporation), of stock of another corporation if, immediately after the acquisition, the acquiring corporation has control of such other corporation (whether or not such acquiring corporation had control immediately before the acquisition);

(C) The acquisition by one corporation, in exchange solely for all or a part of its voting stock (or in exchange solely for all or a part of the voting stock of a corporation that is in control of the acquiring corporation), of substantially all of the properties of another corporation, but in determining whether the exchange is solely for stock the assumption by the acquiring corporation of a liability of the other, or the fact that property acquired is subject to a liability, shall be disregarded;

(D) A transfer by a corporation of all or a part of its assets to another corporation, where the transferor will have control of the corporation to which the assets are transferred;

(E) A recapitalization; or

(F) A mere change in identity, form or place of organization.

From these definitions, it is apparent that the merger of a closely-held company would be an *A, B* or *C* reorganization.

Where the seller receives stock in the acquiring company, the question arises as to how long he must keep it. Insofar as the IRS is concerned, the form of the merger has a bearing on tax aspects, and the stockholder should check with his CPA to determine the impact of taxes on the sale of his stock. The holding period is likely to be determined by restrictions of the acquiring company, however, since they do not want the stockholder to dump sizable blocks of stock and cause a possible drop in its market price.

SECURITIES AND EXCHANGE COMMISSION

The speculative and unethical practices that preceded the stock market crash of 1929 led to two major pieces of corrective legislation. The first was the Securities Act of 1933, designed to protect investors by requiring full and accurate information regarding securities, and to prevent deceit or other fraudulent practices in the sale of securities to be offered. One provision of the act states that a security cannot be offered for sale until a registration statement has been filed with the Securities and Exchange Commission (SEC). This statement must include a prospectus to be provided to each purchaser of the security.

The second act was the Securities Exchange Act of 1934, which concerns trading on national exchanges and on the over-the-counter market, after the initial distribution of securities. Various sections of the act require the filing of annual, semi-annual and other reports; prevent insiders from profiting on inside information; prohibit fraudulent devices and practices; and require a registration statement essentially similar to that required under the 1933 act.

With regard to mergers, the SEC has taken the position that a merger is primarily a corporate act under state law rather than an offering of a security for sale. Hence the issuance of securities does not require registration under the Securities

Act; but there are specific limitations to this exemption. To comply with SEC requirements, an acquirer who offers securities to the sellers of a closely-held company must ensure that the individuals receiving stock are not merely channels for a wider, public distribution of the stock, since in this case the private-offering exemption would not be available. For this reason the issuer frequently requires "investment letters" from those acquiring the stock, stating that they are receiving the stock for investment and not for redistribution. To ensure compliance with this intent, the issuer may also put a restrictive statement on the stock certificate and give "stop transfer" directions to the transfer agent.

Another provision of the 1934 act requires an acquiring company to report any "material" acquisition; that is, one whose assets or gross revenues are 15 per cent or more of those of the acquiring company.

APPENDIX C
BIBLIOGRAPHY

This bibliography is intended for the businessman who does not have time to look through a mass of reading material for additional information on specific aspects of mergers. Toward this purpose, the references are keyed to the sections of this book, so that the reader who wants to read more about valuation, for instance, can say to his secretary, "Miss Jones, get me the references I've circled here under column three."

The fact that some references have not been checked does not imply they are not worthwhile reading; it means that they may pertain to aspects of mergers other than selling a company, such as the acquirer's viewpoint, or merger history or social aspects of mergers.

BOOKS

Book	1. Overview	2. Why Sell?	3. Valuation	4. Inside the Buyer's Tent	5. Earn-Out	6. Selecting the Right Buyer	7. Increasing Company Value	8. Merger Broker	9. Legal, Accounting Specialists	10. Form	11. Negotiations	12. Contract and Closing	13. After Merger	App. B Regulatory Aspects
Alberts, William W., and Segall, Joel E. *The Corporate Merger.* Chicago and London: The University of Chicago Press, 1966.		●			●	●						●		
Bock, Betty. *Mergers and Markets: an Economic Analysis of the First Fifteen Years Under the Merger Act of 1950.* New York: National Industrial Conference Board, 1966.													●	
Bosland, Chelcie C. *Estate Tax Valuation in the Sale or Merger of Small Firms.* New York: Simmons-Boardman, 1963.	●	●												
Butters, J. Keith, Lintner, John, and Cary, William L. *Effects of Taxation, Corporate Mergers.* Boston: Graduate School of Business Administration, Harvard University, 1951.	●													
Choka, Allen D. *Buying, Selling and Merging Businesses.* Philadelphia: Joint Committee on Continuing Legal Education of the American Law Institute and the American Bar Association, 1965.									●				●	
Dellenbarger, Lynn E. *Common Stock Valuation in Industrial Mergers.* Gainesville, Fla.: University of Florida Press, 1966.					●									
Finance Division, American Management Association. *Corporate Mergers and Acquisitions.* New York: American Management Assn., 1958.	●	●	●											
Hennessy, J. H., Jr. *Acquiring and Merging Businesses.* Englewood Cliffs, N.J.: Prentice-Hall, 1966.	●	●	●				●	●				●	●	
Hutchison, G. Scott (editor). *The Business of Acquisitions and Mergers.* New York: Presidents Publishing House, 1968.		●	●					●				●		
Kelly, Eamon M. *The Profitability of Growth Through Mergers.* University Park, Pa.: Pennsylvania State University, 1967.			●									●		
Linowes, David F. *Managing Growth Through Acquisition.* New York: American Management Association, 1968.		●	●											

BOOKS

Source	1. Overview	2. Why Sell?	3. Valuation	4. Inside the Buyer's Tent	5. Earn-Out	6. Selecting the Right Buyer	7. Increasing Company Value	8. Merger Broker	9. Legal, Accounting Specialists	10. Form	11. Negotiations	12. Contract and Closing	13. After Merger	App. B Regulatory Aspects
Little, Arthur D., Inc. (Drayton, Clarence I., Jr., Emerson, John, and Griswold, John D.). *Mergers and Acquisitions: Planning and Action.* New York: Financial Executives Research Foundation, 1963.		•	•					•	•	•		•	•	
McCarthy, George D. *Acquisitions and Mergers.* New York: Ronald Press, 1963.		•	•					•	•		•	•	•	
Mace, Myles L., and Montgomery, George G., Jr. *Management Problems of Corporate Acquisitions.* Boston: Graduate School of Business Administration, Harvard University, 1962.	•	•	•			•	•		•			•		
Martin, David Dale. *Mergers and the Clayton Act.* Berkeley and Los Angeles: University of California Press, 1959.													•	
Rappaport, Donald. "Buying or Selling a Going Business," *Handbook of Business Administration,* H. B. Maynard (editor). New York: McGraw-Hill, 1967.		•	•						•				•	
Reid, Samuel Richardson. *Mergers, Managers and the Economy.* New York: McGraw-Hill, 1968.														
Scharf, Charles A. *Techniques for Buying, Selling and Merging Businesses.* Englewood Cliffs, N.J.: Prentice-Hall, 1964.		•	•		•		•		•		•			
Short, Robert A. *Business Mergers—How and When to Transact Them.* Englewood Cliffs, N.J.: Prentice-Hall, 1967.														
Wintrub, Warren G. (Editor). *Planning Business Combinations.* Lybrand, Ross Bros. & Montgomery, 1968.		•	•					•					•	

PERIODICALS

Source	1. Overview	2. Why Sell?	3. Valuation	4. Inside the Buyer's Tent	5. Earn-Out	6. Selecting the Right Buyer	7. Increasing Company Value	8. Merger Broker	9. Legal, Accounting Specialists	10. Form	11. Negotiations	12. Contract and Closing	13. After Merger	App. B Regulatory Aspects
"After the Merger: Firms That Combine Often Find the Period of Adjustment Painful," *Wall Street Journal,* March 13, 1967.	•											•		
"At Last, Merger Guidelines for Merger-Minded U.S. Companies," *Newsweek,* June 10, 1968, p. 68.													•	

PERIODICALS

Reference	1. Overview	2. Why Sell?	3. Valuation	4. Inside the Buyer's Tent	5. Earn-Out	6. Selecting the Right Buyer	7. Increasing Company Value	8. Merger Broker	9. Legal, Accounting Specialists	10. Form	11. Negotiations	12. Contract and Closing	13. After Merger	App. B Regulatory Aspects
"Biggest, Wildest Year Ever," *Fortune*, June 15, 1968, pp. 43–44.														
Bradley, J. F. "Mergers: A Systems Approach," *Duns Review*, April 1968, pp. 19–20.														
Butler, John J. "Maximizing Personnel Potential In a Merger," *Mergers and Acquisitions, the Journal of Corporate Venture*, Fall, 1965, pp. 74–77.					•							•		
Campobello, Richard, and Loscocco, S. John. "When To and How To . . . In a Merger or Acquisition," *Mergers and Acquisitions, the Journal of Corporate Venture*, Spring, 1967, pp. 65–80.		•	•											
Chambers, Robert L. "How Not to Sell Your Company," *Harvard Business Review*, May–June 1961, pp. 105–108.	•				•	•						•		
"Conglomerates: The Merger-Minded Wonders of Wall Street," *Changing Times*, the Kiplinger Magazine, April 1968, pp. 25–29.	•				•									
D'Aleo, Joseph S. "The Con Game," *Mergers and Acquisitions, the Journal of Corporate Venture*, July–August 1968, pp. 30–33.					•									
D'Aleo, Joseph S. "The Conglomerates: A Wall Street Reappraisal," *Mergers and Acquisitions, the Journal of Corporate Venture*, May–June 1968, pp. 48–53.					•									
Davis, R. E. "Compatibility In Corporate Marriages," *Harvard Business Review*, July–August 1968, pp. 86–93.					•							•		
Depew, Samuel. "A Reader's Guide to Mergers and Acquisitions," *Mergers and Acquisitions, the Journal of Corporate Venture*, Winter, 1967, pp. 58–73.	•													
Elliott, Robert M. "An SEC Primer," *Mergers and Acquisitions, the Journal of Corporate Venture*, Spring, 1966, pp. 55–63.													•	
"Give Us a Merger Policy," *Dun's Review*, December 1967, p. 43.												•	•	

PERIODICALS

Periodical	1. Overview	2. Why Sell?	3. Valuation	4. Inside the Buyer's Tent	5. Earn-Out	6. Selecting the Right Buyer	7. Increasing Company Value	8. Merger Broker	9. Legal, Accounting Specialists	10. Form	11. Negotiations	12. Contract and Closing	13. After Merger	App. B Regulatory Aspects
Grimm, Willard T. "A Banker's Role In Corporate Merger Planning," *Banking*, August 1966, pp. 45–46.									•					
Hecht, Charles J. "Earn-Outs," *Mergers and Acquisitions, the Journal of Corporate Venture*, Summer, 1967, pp. 2–16.					•									
Hemming, Noel P. "Merger Minded Stocks," *Mergers and Acquisitions, the Journal of Corporate Venture*, July-August 1968, pp. 72–73.						•								
Hexter, Richard M. "How to Sell Your Company," *Harvard Business Review*, September-October 1968, pp. 71–77.	•		•			•								
"How FTC Keeps Up On Mergers," *Business Week*, May 25, 1968, pp. 132–133.													•	
Kilmer, David C. "Growth by Acquisition: Some Guidelines for Success," *Business Horizons*, Spring, 1967, pp. 55–62.				•						•		•		
Kitching, J. "Why Do Mergers Miscarry?" *Harvard Business Review*, November-December 1967, pp. 84–101.				•								•		
"Learning the Ways of Matchmaking," *Business Week*, April 13, 1968, pp. 127–128.														
Linowes, David F. "The CPA's Role in Mergers," *Management Services*, September-October 1967, pp. 49–52.									•					
Liston, Robert A. "Securities Fraud and the S.E.C.," *Mergers and Acquisitions, the Journal of Corporate Venture*, May-June 1968, pp. 40–48.													•	
Markstein, David L. "Relative P/E Ratio—Newest Statistical Tool In Merging," *Mergers and Acquisitions, the Journal of Corporate Venture*, March-April 1968, pp. 11–17.						•								
Markstein, David L. "What Happens to the Stock?" *Mergers and Acquisitions, the Journal of Corporate Venture*, Winter, 1966, p. 50.						•								
"Marriage Brokers; Merger Brokers," *Time*, November 19, 1965, p. 109.								•						

PERIODICALS

Citation	1. Overview	2. Why Sell?	3. Valuation	4. Inside the Buyer's Tent	5. Earn-Out	6. Selecting the Right Buyer	7. Increasing Company Value	8. Merger Broker	9. Legal, Accounting Specialists	10. Form	11. Negotiations	12. Contract and Closing	13. After Merger	App. B Regulatory Aspects
"Measuring the Surge of Mergers," *Business Week*, March 23, 1968, p. 69.														●
"Merger of the TRW/STL Product Engineering Laboratory with the TRW Pacific Semiconductor, Inc., Integrated Circuit Group," *Mergers and Acquisitions, the Journal of Corporate Venture*, Fall, 1966, pp. 34–35.									●		●	●		
"Mergers on Parade," *Mergers and Acquisitions, the Journal of Corporate Venture*, January-February 1968, pp. 86–87.			●		●									
Morgello, C. "Wall Street: the Merger Market," *Newsweek*, Dec. 2, 1968, p. 76.		●		●										
"Multicompanies: Conglomerate, Agglomerate, and In-Between; Multi-Industry; With Yardsticks of Management Performance," *Forbes*, Jan. 1, 1969, pp. 77–86.					●									
O'Toole, Edward T. "Marko's Minimum-Risk Merger Method," *Mergers and Acquisitions, the Journal of Corporate Venture*, Winter, 1967, pp. 55–57.				●										
O'Toole, Edward T. "Merger Master: Salgo, Financial Engineer," *Mergers and Acquisitions, the Journal of Corporate Venture*, Spring, 1966, pp. 48–54.	●													
Phalon, Richard. "What Wall St. Wants to Know," *Mergers and Acquisitions, the Journal of Corporate Venture*, Winter, 1967, pp. 74–80.		●	●		●									
Poindexter, Joseph. "The Cool, Creative Company Lawyers," *Dun's Review*, March 1969, pp. 35–37.										●				
Pope, Leroy. "To List or Not to List?" *Mergers and Acquisitions, the Journal of Corporate Venture*, January-February 1968, pp. 6–16.					●									
"Probing the New Giants; FTC Investigation of Conglomerates," *Business Week*, July 13, 1968, p. 37.													●	
Reed, Stanley Foster. "The Merger Boom—How Long Can It Last?," *Mergers and Acquisitions, the Journal of Corporate Venture*, Spring, 1967, pp. 8–9.					●									

PERIODICALS

Periodical	1. Overview	2. Why Sell?	3. Valuation	4. Inside the Buyer's Tent	5. Earn-Out	6. Selecting the Right Buyer	7. Increasing Company Value	8. Merger Broker	9. Legal, Accounting Specialists	10. Form	11. Negotiations	12. Contract and Closing	13. After Merger	App. B Regulatory Aspects
Reed, Stanley Foster. "Primer for a President," Part One, *Mergers and Acquisitions, the Journal of Corporate Venture,* Fall, 1965, pp. 21–25.			●					●	●					
Reed, Stanley Foster. "Primer for a President," Part Two. *Mergers and Acquisitions, the Journal of Corporate Venture,* Winter, 1966, pp. 67–73.			●											
Rockwell, Willard F., Jr. "How to Acquire a Company," *Harvard Business Review,* September-October 1968, pp. 121–132.	●		●			●						●	●	
"Runaway Boom in Mergers; Crackdown Coming?" *U.S. News and World Report,* July 29, 1968, pp. 61–62.													●	
"Sardine That Became a Whale," *Fortune,* June 15, 1968, p. 260.														
Simons, M. "Stock Trends," *Forbes,* Oct. 15, 1968, p. 104.						●								
"Thinning the Ranks; Dismissal of Kelvinator and Norge Personnel After Mergers," *Business Week,* July 27, 1968, p. 36.														
"To The Firing Line," *Mergers and Acquisitions, the Journal of Corporate Venture,* July-August 1968, pp. 68–71.														
Weiner, J. B. "Anatomy of an Acquisition," *Dun's Review,* May 1966, p. 34.			●											
Wendel, W. H. "Is There a Perfect Merger?" *Dun's Review,* April 1967, pp. 37–38.												●	●	
"What Little Business Thinks About Mergers," *Fortune,* August 1967, p. 94.														
"What to Do About a Merger," *Dun's Review,* February 1968, pp. 101–102.			●											
"Why Corporate Merger Plans Go On the Rocks," *Business Week,* November 23, 1968, p. 124.														
Zwerdling, George H. "In Defense of Conglomerates," *Mergers and Acquisitions, the Journal of Corporate Venture,* May-June 1968, pp. 54–57.						●								

The following terms are defined in their customary business usage as pertains to mergers. The meanings may differ slightly therefore from their use in technical or dictionary terminology.

Boot—Cash or other property received, in addition to voting securities, as consideration in a merger. If substantial boot is received, the transaction may be taxable rather than a tax-free exchange.

Buy and Sell Agreement—The businessmen's agreement covering price, basic terms, whether the transaction is an exchange of stock or sale of assets, and other basic matters in the sale or merger. This agreement does not go into legal matters. It is signed at the earliest feasible time, sometimes before attorneys are brought into the discussions.

"Chinese Money"—Common-stock selling at an extremely high price/earnings ratio, sometimes 60, 70 or 100 times current earnings. An acquiring company can use its own stock for acquisitions and can pay a much higher price in stock than it can in cash; for this reason the stock is referred to as "Chinese money" by the financial community. Reported earnings per share sometimes do not take into account the dilution caused by conversion of other securities into common stock, or the effect of exercising warrants.

Closing—The transaction completing the merger. This is a meeting of the principals, their attorneys and other counsel at which time the seller delivers to the buyer such items as stock certificates, stock book, lawyers' opinions, assignment of patents, certificate of warranties and representations, minute books and copies of various agreements between the two parties. The buyer delivers to the seller his certified check or notes, stock or other securities, copies of resolutions authorizing the transaction, opinion of counsel as to corporate legal standing, the buyer's representations and warranties and other items related to the merger.

Conglomerate—A company that has achieved wide diversification, primarily through acquisitions. Most conglomerates prefer to be called "multimarket companies," "widely-diversified companies" or "acquisition-minded companies."

Deferred Tax Payment Plan—See *Installment Sale.*

Earn-Out—An agreement between buyer and seller whereby the price received by the seller may increase (or decrease) according to whether the acquired subsidiary's profits measure up to expected results. The earn-out formula generally is based on profit performance expected by the seller. The earn-out payment usually is in the form of stock in the acquiring company.

"Funny Money"—Warrants (options to buy common stock at specified prices) used by acquisition-minded companies to acquire other companies. The term is also sometimes used interchangeably with "Chinese money."

Holding Company—A corporation that owns the stock or assets of one or more operating companies and provides common ownership and overall financial management, but does not enter into day-to-day operations as does an operating company.

Horizontal Integration—The acquisition of a company in the same general business as the buyer, but in perhaps a different location or on an expanded scale. An electric motor manufacturer in California who acquires a similar business in Ohio would be making a horizontal integration acquisition.

Installment Sale—As applied to the sale of a business, taxes on gains may be paid in installments corresponding to payments received from the buyer, provided not more than 30 per cent of the selling price is received in the year in which the transaction is completed. (Most attorneys and accountants try to keep the payment at 29 per cent.) If the initial payment and subsequent monthly or quarterly payments in the year of the sale exceed 30 per cent, the transaction is subject to full taxation in the year of the sale.

Leverage—The utilization of debt financing to increase return on equity. For example, a company that sells $500,000 in stock (equity) and borrows $500,000 would use the entire $1 million to produce profit, but would show a return on the $500,000 equity (less cost of borrowing).

Long-Term Capital Gains—The selling price of a company less the owner's original investment in it. Federal capital gains tax on a company (or security) held for more than six months is generally 25 per cent of the profit realized from the sale.

Merger—A combination of two companies with one company surviving. The selling company is usually merged into the acquiring company. Basically there are three types of tax-free mergers, as provided in Section 368 of the IRS Code: *A*—statutory merger; *B*—stock-for-stock exchange; *C*—stock-for-assets exchange.

Merger Broker—A professional businessman authorized by the seller (or buyer) to find the desired counterpart. He helps the seller analyze and prepare his company for merger, negotiates on behalf of his client and assists in integration of the two firms. He may also be referred to as a merger packager, merger specialist or merger intermediary.

Earnings per share—The after-tax earnings divided by the number of shares outstanding.

Pooling—For accounting purposes, the acquisition of one company by another is treated as a "purchase" or a "pooling of interests." In a pooling, where the consideration usually is voting equities, ownership of each enterprise is considered as having always been in a combined state. No new basis for accounting is required, and accounting balances and historical data are combined (see *Purchase*).

Price/Earnings Multiple—Most closely-held companies are not listed on major exchanges (although they may be traded over-the-counter) and therefore do not have an established price/earnings ratio (see below). The selling price of a closely-held company can be computed as a multiple of its after-tax earnings; the selling price divided by after-tax earnings is the price/earnings multiple.

Price/Earnings Ratio—Market price of a traded stock divided by the earnings per share. A company that has 500,000 shares of stock issued and outstanding and whose after-tax earnings are $1 million would have an earnings per share of $2. If its stock sells at $30 per share, it would have a price/earnings ratio of 15:1. Earnings per share would be diluted, and the price/earnings ratio higher, if warrants and options are converted into stock. Some companies now report earnings per share on the basis of issued and outstanding shares, and also on a fully-converted basis.

Purchase—In purchase accounting, the price paid often exceeds the value of specific assets and results in payment for goodwill. Payment is often made in cash or securities. Because of the adverse effect (to the acquirer) of accounting and taxes in a purchase, he will often offer less to the seller than he would in pooling of interests accounting.

Tax-Free Merger—In reality, a tax-deferred transaction whereby the seller of a company may defer tax on capital gains if the transaction complies with IRS regulations. In a stock-for-stock exchange, the seller generally would not be taxed until he sells stock of the acquiring company received in the transaction.

Triangular Merger—The term applied where a Type *A* merger (see *Merger*) is effected by the buyer's setting up a wholly-owned subsidiary into which the seller's company is merged. The board of the acquiring company votes all of the stock in the subsidiary, avoiding the delays and expense of the acquiring company's having to obtain shareholder approval.

Warrants—See *"Funny Money."*

Vertical Integration—The acquisition of a company where the products or services of one corporation (buyer or seller) become part of the product or services of the other. A manufacturer of electric motors who acquires a foundry to produce castings for the motors would be making a vertical integration acquisition.

INDEX

A

Accountants, 118–119
  fees paid to, 143
Accounting
  adjustments in, 94
  effect on valuation of, 37
Acquisition Agreement, 156–57
Acquisitions, 39–54
  financing of, 48–49
  pitfalls in, 46–48
  process of, 41–45
  reasons for, 39–41
Adjustments
  accounting, 94
  for savings, 33
Alco Standard Corporation, 49
American Home Products Corporation, 99
American Stock Exchange (Amex), 74
Analysis
  financial, 32
  by merger broker, 104
Anheuser-Busch, Inc., 74
"Annual average" profit-unit earn-out, 60
Appearance vallue, 80–82
Appraisal formulas, 35
Arthur Andersen and Company, 94
Assets, sale of, 14
Assets-for-stock exchange, 126–27
Associated Products, Inc., 107
Attorney, 116–18
Audits, 118
Automatic Sprinkler Corporation, 22

B

Bangor Punta Corporation, 40
Bank of America, 98
Bankers, 120

Base period earn-out, 56–58
Basic value, increasing, 82–88
  building on strengths and overcoming
    weaknesses in, 83–88
  profit planning in, 82-83
Book value, 28, 34–35
Boot, 125
Bootstrap operations, 48
Broker, *see* Merger broker
Brokerage contract, 101–4
Brokerage fees, 112, 143
Buchwald, Art, 50
Business agreement, 130–131
  on price, 130
  on terms and form, 130–31
Business Week (periodical), 28
Buyer
  representations of, 142
  selection of, 13–14, 65–78, 154–55,
    160–62, 165–66
    "Find Me" approach to, 67–68
    goals and planning in, 76–77
    indirect approach to, 68–69
    knowledge of proposals in, 69–74
    organizational fit and, 75–76
    "Sadie Hawkins" approach to, 66–67
  warranties of, 142
    negotiations on, 132-33

C

CPA, *see* Accountants
Capital, need for, 22
Cash flow, 32
  discounted, 61–63
Certificates of contingent interest, 63–64
Chambers, Robert L., 68
Changing Times (magazine), 70–71
Chase Manhattan Bank, 98

City Investing Company, 64
Clayton Act (1914), 152
Closing, 43, 144–46
    contract clause on, 140
        on conditions of, 142–43
    delays in, 168
Competition
    future earnings and, 32
    as reason for selling, 23–24
Confidential information, 143
Congress, U.S., 171–72
Contingent interest, certificates of, 63–64
Contracts, 139–44
    example of, 140–44
Control
    loss of, 24
    post-merger, 148–49
Corporate charges, 63
Cumulative earn-out, 59

D

Depreciation, 63
Dilution of earnings, 37–38, 71
Discounted cash flow, 61–63
Diversification, 40
    of investments, 21–22
Dividend-paying capability, 29, 36
Documents, access to, 143
Doric Company, 73
Dow Jones Industrial Index, 28
Dun & Bradstreet, Inc., 41
Dyson & Kissner Corporation, 112

E

Earnings
    definition of, 63
    dilution of, 37–38, 71
    future, 31–33
    increased, 40
    past, 31–32
    potential, 28
Earn-outs, 13, 55–64
    certificates of contingent interest and, 63–64
    discounted cash flow and, 61–63
    effect of salary on, 60–61
    formulas for, 56–60
        "annual average" profit-unit, 60
        base period, 56–58
        cumulative, 59
        increment, 58–59
        profit-unit, 59–60
        reverse, 60
Ekco Products Company, 99
Employees, effect of merger on, 149–50

Employment and Competition Agreement, 157
Escape conditions, 133
Escrow agreement, 140, 145
Exhibits, schedule of, 144

F

Federal regulations, 171–74
Federal Trade Commission, 98, 171–72
Fees
    accountants, 143
    brokerage, 112, 143
Financial analysis, 32
Financing of acquisitions, 48–49
Finders, 99–100, 120–22
Fisher Brewing Company, 171
Forbes (magazine), 99–100
Form, 130–31
Former owner, retention of, 150
Future earnings, 31–33

G

Georgia-Pacific Corporation, 49
Glen Alden Corporation, 100
Goodwill, 37
Gross margin, 31
Growth, internal, 71

H

Harris, Upham & Company, 73
Hayes International Corporation, 64
Health Tecna, 73
Hecht, Charles J., 64
Hemming, Noel Phillip, 73
Hoffman, Edward, 112
"How to Acquire a Company" (Rockwell), 75
"How Not to Sell Your Business" (Chambers),
    68
"How to Sell Your Company" (Hexter), 66

I

Identity, loss of, 24
Increased earnings, 40
Increment earn-out, 58–59
Indemnification, 143, 157
Information, access to, 143
Inspection, rights of, 133
Internal growth, 71
Internal Revenue Service, 171, 172–73
    code of, 14, 123–24, 128, 172–73
International Telephone, 73

Investigations
    pre-closing, 43
    preliminary, 41–42
Investment, 155–56
    diversification in, 21–22
Investment advisor, 119–20
Issues, recording of, 134–35

J

Justice Department, U.S., 98
    Antitrust Division of, 171–72

K

Kelly, Eamon M., 28n

L

Lawyer, 116–18
Lear Siegler, 73
Legal counsel, 116–18
Legal negotiations, 131–33
    on buyer warranties, 132–33
    on escape conditions, 133
    on rights of inspections, 133
    on seller warranties, 132
Lehman Brothers, 99
Ling-Temco-Vought, 73
Lucky Lager, 171

M

Mace, Myles L., 25, 94
"Making of a Conglomerate—a Buchwald
    Fantasy" (Buchwald), 50
Management consultant, 120
Margin, gross, 31
Market, future earnings and, 32
Market prices of stocks, 29, 35
Marko, Harold, 60
Maxad Corporation, 73
Merger broker, 97–114
    fees of, 112
    functions of, 100–12
        analysis, 104
        brokerage contract, 101–4
        negotiations, 107–12
        post-merger integration, 112
        search, 104–7
Mergers and Acquisitions (periodical), 64
Method of payment, 36–37
Mid-Continent Telephone Corporation, 49

Montgomery, George C., Jr., 20, 94
Moody's Investors Service, Inc., 41, 71
Mueller Steam Specialty Company, 61–63

N

Negotiations, 43, 129–38, 166–67, 169–70
    business agreement in, 130–31
        on price, 130
        on terms and form, 130–31
    human factor in, 135–36
    in larger companies, 136–38
        organizing for, 136
        strategy and tactics for, 137–38
    legal, 131–33
        on buyer warranties, 132–33
        on escape conditions, 133
        on rights of inspections, 133
        on seller warranties, 132
    merger broker in, 107–12
    preparation for, 133–34
        role-playing in, 134
    recording of issues in, 134–35
    timing of sessions for, 135
    written agreements in, 135
Negotiator, attributes of good, 136–37
New York Stock Exchange, 74
North American Rockwell Corporation, 29, 67

O

Objectives, determination of, 41
Offer
    attractive, 24
    unacceptable, 25
Omega Equities Corporation, 49
Operations, post-merger, 147–48
Organizational fit, 75–76
Over-the-counter (OTC) stocks, 74–75
Owner, retention of former, 150

P

Palevsky, Max, 82
Past earnings, 31–32
Patent rights, 29, 36
Payment
    method of, 36–37
    of purchase price, 140
Performance, 32
Personnel as valuation factor, 29
Planning
    post-merger, 148–49
    profit, 82–83
    in selection of buyer, 76–77
Plant as valuation factor, 30
Pooling of interests, 14, 37

Position gap, 23
Post-merger conditions, 147–51
   effect on employees of, 149–50
   former owner retained in, 150
   of operations, 147–48
   planning and controls in, 148–49
Post-merger integration, 45, 112
Potential as valuation factor, 30–31
Pre-closing investigation, 43
Preliminary investigation, 41–42
Preliminary screening, 41
Price, 130
   limits on, 37–38
   purchase, 140
   See also Valuation
Price/earnings multiple, 33–34
Price factors, 12–13
Product development, 22–23
Product as valuation factor, 30
Profit
   planning of, 82–83
   as valuation factor, 31
Profit-unit earn-out, 59–60
Promissory Note, 156
Purchase, 14, 37
Purchase and Sell agreement, 14–15
   legal form of, 139–40
   signing of, 144
Purchase price, 140

R

Recording of issues, 134–35
Registration Agreement, 157
Regulations, federal, 171–74
Release and Assignment, 157
Representations
   of buyer, 142
   of seller, 141
   survival of, 143
Retirement, lack of succession and, 23
Reverse earn-out, 60
Rights of inspection, 133
Rockwell, Willard F., Jr., 29
Rockwell Manufacturing Company, 51, 74
Role-playing, 134
Royal Industries, 73

S

Salaries, 63
   effect on earn-out of, 60–61
Sale of assets, 14
Savings, adjustments for, 33
Scientific Data Systems (SDS), 82–83
Schedule of exhibits, 144
Schenley Industries, Inc., 100

Screening, preliminary, 41
Search, 104–7
Securities, 70–74
Securities Act (1933), 173
Securities and Exchange Commission, 171,
   173–74
Securities Exchange Act (1934), 173–74
Security Pacific Bank, 98
Seller
   representations of, 141
   warranties of, 141–42
      negotiations on, 132
Selling, 19–26
   decision on, 25–26, 164–65
   negative factors, in 24–25
   reasons for, 20–24
      attractive offer, 24
      competition, 23–24
      investment diversification, 21–22
      need for capital, 22
      product development, 22–23
      retirement and lack of succession, 23
      silent partner situation, 24
      tax considerations, 21
Shares, sale and purchase of, 140
Sherman Antitrust Act (1890), 171
Silent partner, 24
Small Business Administration, 162
Small business qualification, loss of, 25
Soss Manufacturing Company, 60, 61–63
Spacek, Leonard, 94
Specialists, 115–22
   bringing in, 122
   roles of
      accountant, 118–19
      investment advisor, 119–20
      legal counsel, 116–18
   selection of, 115–16
Specialized service, 25
Statutory merger, 14
   tax-free status of, 124–25
Standard & Poor's Corporation, 41, 71
Stock Purchase Agreement, 140–46
Stock-for-stock exchange, 14
   tax-free status of, 125–26
Stockholders' Letter of Agreement, 157
Stocks
   market price of, 29, 35
   over-the-counter (OTC), 74–75
Succession, lack of, 23
Supreme Court, U.S., 172
Survival of representations and warranties, 143

T

Taxable transactions, 127–28
Taxes, 63
   as reason for selling, 21

Tax-free transactions, 123–27
    assets-for-stock exchanges as, 126–27
    statutory mergers as, 124–25
    stock-for-stock exchanges as, 125–26
Tax-loss carry-over value, 36
Technology, 40
Teledyne, 73
Terms, 130–31
Textron Corporation, 73
Transactions, 14–15, 123–28
    major documents in, 156–57
    smooth, 170
    taxable, 127–28
    tax-free, 123–27
TRW Company, 73

U

Unacceptable offer, 25
U.S. Industries, Inc., 49

V

Valuation, 27–38
    effect of accounting aspects on, 37
    effect of method of payment on, 36–37
    factors in, 28–36
        appraisal formulas, 35
        book value, 34–35
        future earnings, 31–33
        personnel, 29
        plant, 30
        potential, 30–31
        price/earnings multiple, 33–34
        product, 30
        profit, 31
    limits on buyer's price and, 37–38

Value, 79–95
    appearance, 80–82
    increasing basic, 82–88
        building on strengths and overcoming
            weaknesses in, 83–88
        profit planning in, 82–83
    knowledge of, 94–95

W

Walter Kidde & Company, 49
Warranties
    buyer, 142
        negotiations on, 132–33
    seller, 141–42
        negotiations on, 132
    survival of, 143
Weighted average, 32
Wintrub, Warren G., 77n
Written agreements, 135

X

Xerox Corporation, 82

Z

Zero Manufacturing Company, 73